THINKING METRIC

Thomas F. Gilbert

and

Marilyn B. Gilbert

Praxis Corporation
New York, New York

John Wiley & Sons, Inc.

New York · London · Sydney · Toronto

Library of Congress Cataloging in Publication Data

Gilbert, Thomas F
    Thinking metric.

    1. Metric system.   2. Weights and measures--
United States.   3. Weights and measures--Canada.
I. Gilbert, Marilyn B., joint author. II. Title.
QC91.G55   389'.152   73-6997
ISBN 0-471-29901-4

Printed in the United States of America

73 74 10 9 8 7 6 5 4 3 2 1

To the vanishing gilbert

Preface

When we first decided to write this book, we thought we would have to be
very selective about our audience. After all, the machinist and the cook
make a totally different use of measurements. But we soon realized that
all of us are up against the same thing. All of us must learn to speak
and understand a new language fluently; and most of us are beginning from
ground zero, regardless of how close to measurement our jobs may take
us. Once this was clear, our task became clear, too, and somewhat
easier. First, we obviously could write the same book for the machinist
as for the cook. And second, we saw that we must all learn the new language directly, not through translation.

This book includes many problems and exercises that give practice
in using this new language. Not every reader will want—or even need—
all that practice. Each reader's common sense (or natural laziness)
can make the decision of how much practice is enough.

Because some readers may want to understand more about the quantities being measured, we have included some discussions of temperature,
weight, power, and energy. But readers who don't need these elementary
discussions can quickly find their way around them. (Indeed, by referring
to the guide on how to use this book, many readers will find their own way
through these lessons in an hour or two.) In any case, our discussions
of physics are very elementary, and we make no claim to teach physics.

Nor do we attempt to teach arithmetic, except for reviewing the use
of exponents of 10 for students who might use the conversion tables in
the Appendix. We assume that readers have mastered elementary arithmetic. The arithmetic we do use isn't difficult. In fact, one of the
beauties of the metric system is that it eliminates fractions and mixed
numbers. What else was hard about elementary arithmetic?

Those instructors who want to assign the book to students can indicate
which parts of the metric system are most important to their courses.
This would give their students guidance in how carefully and extensively
to study the different units.

The first part of Chapter 8 is very important; it will help readers
maintain their mastery in thinking metric. The rules-of-thumb for
conversions from the customary system to the metric system are tools
for practicing these new metric skills on a day-to-day basis.

The first person we want to thank is Duke Schmidt of the Ford Motor Company. Duke was responsible for convincing us that we should do the book. We also thank our reviewers, particularly Dan Leahy of John Jay College, whose help is no excuse for any errors we have made in physics. We are especially grateful to Irene Brownstone of John Wiley & Sons, that artist among editors, and Judy Wilson, also of John Wiley & Sons, who helped us in too many ways to measure. Thanks and many kisses to Robby Gilbert and Eve Gilbert, whose cooperation was essential.

New York   Thomas F. Gilbert
September 1973   Marilyn B. Gilbert

How To Use This Book

This book is self-instructional; you should be able to tailor it to your own needs. How you should use this book really depends on your experience and your inclinations. If you study the table below and identify the kind of reader you are, you can mark the pages that are relevant for you.

If you are this kind of reader . . .	. . . read these sections:

I.   (a) Have a fair idea of what the metric system is all about
    (b) Not a student or employee of science or technology
    (c) Not greatly fond of reading
    (d) No time to waste
	Chapter 2, pages 11–17
	Chapter 3, pages 19–25, 29–30, 32–34
	Chapter 4, pages 44–46
	Chapter 5, pages 53–60
	Chapter 6, pages 66–69
	Chapter 8, pages 87–99
	
	Estimated time: 1 to 3 hours

II. Same as I, except a student or employee of science or technology	Same as I, but add Chapter 7. Estimated time: 2 to 4 hours
III. Same as I, except in no hurry and enjoy reading	Read the whole book except Chapter 7 and Part B of Chapter 8. Estimated time: 3 to 6 hours
IV. Same as II, except in no hurry and enjoy reading	Read the whole book. Estimated time: 4 to 7 hours
V. Know very little about the metric system.	Read the whole book. Estimated time: 4 to 7 hours

    The Self-Tests (page 105) will help you evaluate your learning. The questions are grouped by chapter, so you may take the appropriate test after reading each chapter or you may take all the Self-Tests at once after you have completed the book.

You may also use these Self-Tests as previews of the material covered in each chapter. If you can easily answer the questions for a chapter, you can either skip the chapter or just skim it lightly.

For those who need to make precision conversions now, before we go completely metric, we have included conversion tables in the Appendix. We have also provided a list of materials for further reading.

Contents

THINKING METRIC

CHAPTER ONE

Why Think Metric?

Gone is the mile of measured length
And pint and pound and peck;
Gone the gilbert of magnetic strength,
The dram no more correct.

And though the inch-worm does not change
As the inch is banished,
Won't the earth seem wondrous strange
Now the foot has vanished?

The United States and Canada are in the middle of a conversion from
the system of measuring in feet and pounds to another mode of measurement called the metric system.   This means that

. . . the inch ruler will give way to the centimeter (pronounced SENTA-
meter) ruler . . .

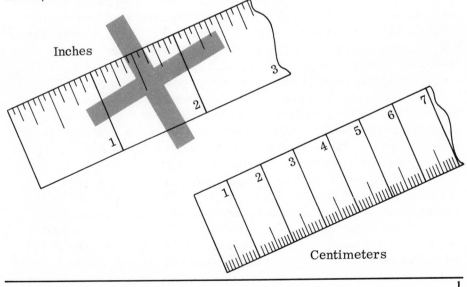

. . . the ounce will yield to the milliliter (pronounced MIL-y-leeter) . . .

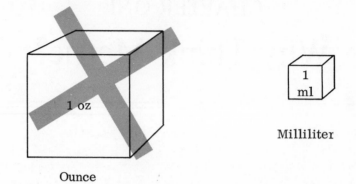

Ounce

Milliliter

. . . and the pound will pass to the kilogram (pronounced KILL-a-gram).

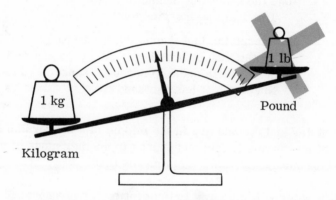

Kilogram

Pound

Very shortly, the weatherman will be predicting a nice day in spring as 20° Celsius (SELL-shuss) rather than 68° Fahrenheit. When you buy a car, you will buy kilowatts instead of horsepower. Your air conditioner will be rated in joules (JOOLS) rather than Btu's. When you study physics in school, you will compute newtons of force instead of pounds, and pascals (PASS culls) of pressure instead of pounds per square inch.

We first decided to write this book because all the evidence showed that the United States would follow everyone else in the world and convert to the metric system of measurement. (Canada's conversion program is now official.) But as we did our research, we discovered—much to our surprise—that metric conversion is not a promise of the near future; it has actually been happening for quite a few years. Many industries are already largely metric. For example, a Bayer aspirin tablet is exactly 1 centimeter in diameter. Skis are sold or rented in metric sizes. The circumference of a common deodorant bottle is in metric measures. The list is very long indeed. More important, the United States Senate has already passed a bill that will convert the government to a metric basis. Those companies selling to the government will have to supply metrically measured products. And there will be meters, not miles, of red tape.

Metric conversion will soon become public. Congressional and administrative leaders have made it clear that the question is no longer, "Will all the North American people buy their butter in grams rather than ounces and pounds?" They will. Nor is the question, "When will this conversion happen?" It is already happening. The only remaining question is, "How will this conversion be completed?" Many major corporations have established programs to complete their conversion. Among them are Ford Motor Company, General Motors, IBM, Honeywell, and Caterpillar. But the main problem in completing this conversion is primarily one of public education. How can the people of North America be taught to think in meters instead of feet, in milliliters instead of ounces, and in kilograms instead of pounds?

This book is designed to help you to think in the metric system—even better than you think in the customary North American system (NAS)—after an investment of only a few hours.

Why are we converting to the metric system? First, the metric system is much simpler than the NAS system of feet, pounds, gallons, and degrees Fahrenheit. This should be reason enough. But there is another important reason for adopting the metric system: If we don't, soon we won't be able to talk to the rest of the world.

The English language is the most widely spoken language in the world; it is the tongue of diplomacy, international commerce, and science. The prosperity of the English-speaking people has been so great that the world has imitated us in every area of life—our foods, our fashions, and our follies from pop music to soda pop. The principal reason for this position of world eminence has been our achievements in science and engineering. Half the units of science that honor a man have English names: Joule, Newton, Watt, Gilbert, Henry, Faraday, and Maxwell are examples. In fact, in the material achievements of science, English-speaking people have had no peers.

If our language, currency, tastes, and styles have become world standards, we would expect that our measurement system would be adopted too—that the world would measure with the foot, the pound, the ounce, the horsepower, and the square mile. But our fundamental language of science and engineering is so clumsy and archaic that we have become isolated provincials.

All the people of the world, except those in Canada and the United States (and a few small countries like Tonga and Liberia), now use the standard, elegant, streamlined language of the metric system. England, Australia, and New Zealand are currently in the middle of a 10-year program of conversion to the metric system. We have been left alone, and at great cost to ourselves.

The cost to industry is one example. Since we buy and sell machinery and parts in the world market, many industries must maintain a double inventory of goods—one in the North American system of measurement and another in the metric system. Converting to the metric system will not only save the North American economy billions of dollars every year, but it will also help us remain competitive in world markets.

The metric system will help us in other ways. One of these—and this is much more important than we might at first think—will be in simplifying the basic arithmetic and science taught in schools. As a conservative estimate, 15 percent of the time spent in elementary arithmetic is used to teach such tiresome skills as finding the least common denominator, reducing improper fractions, adding mixed numbers, and reducing fractions to their simplest terms. Since these skills are primarily needed for performing arithmetic using feet and inches and pounds and ounces, they presumably can be eliminated from the curriculum. The time spent in teaching just those skills may cost us well over a billion dollars a year.

This relief will not be limited to elementary school students. About 20 percent of every arithmetic textbook for nurses is devoted to these clumsy calculations. Why arithmetic calculations in college? Because they are so difficult to learn that many students fail to master them in the lower grades. Nurses may actually need to make such conversions —or at least understand them—because dosages of drugs are often measured in one measuring system and administered in another. Yes, the metric system will make life simpler for everyone.

In 1960, representatives of most of the important international scientific organizations agreed to adopt a simplified form of the metric system for scientific use throughout the world. This simplified metric system is called SI (an abbreviation for Systéme Internationale d'Unitès), or, in English, the International System of Units.* The SI version is simpler than the former metric system because it discards a number of metric units that aren't absolutely necessary (e.g., the calorie as a measure of energy and the gilbert as a measure of magnetomotive force).

In this book, you'll learn the most common units of the simplified SI version of the metric system just as you should learn any new language —by actually using the words in a practical context. To think metric just as naturally as you think in your native language, you have to practice. The exercises will give you that opportunity.

You will not only "think metric," but you will probably think far better in the metric system than you have ever thought in our own North American system. (Can you say now how much a nickel weighs, for example, or estimate the number of square feet in a room?) After you

---

* In 1972, the international agreement was reaffirmed. The United States and Canada signed the agreement along with the other major industrial nations of the world.

have worked the problems here, you will be able to conceptualize distances, weights, volumes, circumferences, energy, power, pressure, and force.

You needn't wait for your country to complete its conversion to the metric system before you put this system to practical use. You will find many of these units useful to you right now—particularly if you study science or nursing, travel in a foreign country, drive a foreign car, or dabble in carpentry. But even if you don't do any of these things, you will need to know what these metric units are so you can understand the articles you'll be seeing more and more frequently in newspapers and magazines as we go completely metric.

CHAPTER TWO
Language of the Metric System

Watch your p's and q's, and don't resort
To words like "pint" and names like "quart."

Here are two ordinary rulers. If they aren't already familiar, examine them to see how they're different.

Centimeter Scale

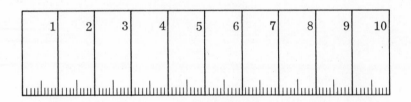

Inch Scale

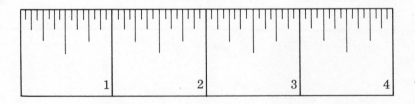

As you see, the marks on the inch ruler are just a little longer than the marks on the centimeter ruler. However, both rulers are equally practical. We could use either to find out how wide a screen we would need for a window, for example, or how much material we would need to cover a chair. But suppose we wanted to adopt just one ruler for everyday use. Would it really make any difference which we chose?

If you look at them closely, you'll note that the inch ruler has 16 marks to the inch whereas the centimeter ruler has 10 marks to the centimeter. And this makes a very great difference indeed. To get an idea of the power of this difference, follow the calculations in this problem:

> Suppose you want tiles to cover your kitchen floor. You measure the length and width of the floor, and then you go shopping. When you find a tile you like, you need to figure out how many of those tiles you will need.

Here is how you would make that calculation, depending on whether you are in a country that uses the North American system of feet and inches or the metric system of centimeters. (A centimeter is abbreviated cm, and a square centimeter is abbreviated cm^2.)

North American System

Problem: How many 8-inch-square tiles will cover a room 13 feet 3 inches by 16 feet 6 inches?

1. Convert feet to inches:

   (a) 13 ft = 13 × 12 in. = 156 in.
   (b) 16 ft = 16 × 12 in. = 192 in.

2. Get the total length and width of the room:

   (a) 156 in. + 3 in. = 159 in.
   (b) 192 in. + 8 in. = 200 in.

3. Find the area of the room:

   159 in. × 201 in. = 31,959 sq in.

4. Find the number of square inches in each tile (the area of the tile):

   8 in. × 8 in. = 64 sq in.

5. Divide the room area by the tile area to get the number of tiles needed:

   31,959 sq in. ÷ 64 sq in. = 499.3 tiles, or 500 tiles

Metric System with SI Units

Problem: How many 20-cm tiles will cover a floor 404 cm by 503 cm?

1. Find the area of the room:

   404 cm × 503 cm = 203,212 cm^2

2. Find the number of square centimeters in each tile:

   20 cm × 20 cm = 400 cm^2

3. Divide the room area by the tile area to get the number of tiles needed:

   203,212 ÷ 400 = 508.03 tiles, or 509 tiles

It's quite clear that the metric system made our calculation easier.
Let's see just how much easier:

Number of Arithmetic Operations	
NAS	METRIC
4 multiplications 2 additions 1 division	2 multiplications 0 additions 1 division

The NAS system has not only required seven operations to the three
needed for the metric system, but two of those extra operations required
a multiplication by twelve—which few of us find easy.

As further proof of the ease of using the metric system, try getting
an estimate of how many tiles you need by doing all the calculations in
your head. Most of us would be unable to do this in the NAS.[*] But in
the metric system, we would reason like this:

1. The area of the room is 404 cm × 504 cm, which is
    roughly 400 cm × 500 cm = 200, 000 cm^2.

2. The area of each tile is 20 cm × 20 cm = 400 cm^2.

3. 400 goes into 200, 000 as 4 goes into 2000, or
    2000 ÷ 4 = 500 tiles needed.

---

* Of course, if you have read Locke's <u>Math Shortcuts</u>, another Wiley Self-
Teaching Guide, you can estimate in the NAS with no difficulty. However, most people don't estimate well. The point is that not only esti-
mations but all calculations are easier with the metric system.

This tile calculation is not an isolated example. Calculations of weight are also simpler with metric units than they are in the NAS. Here is a simple illustration. (The metric measure of weight is the kilogram, which is abbreviated kg.)

North American System	Metric System with SI Units
Problem: Three boxes weigh 7 lb 9 oz, 8 lb 5 oz, and 4 lb 3 oz. What is the total weight in ounces?	**Problem**: Three boxes weigh 3.43 kg, 1.90 kg, and 3.57 kg. What is the total weight?

1. Add the pounds:

   7 lb + 8 lb + 4 lb = 19 lb

2. Convert pounds to ounces:

   19 × 16 oz = 304 oz

3. Add the additional ounces:

   9 oz + 5 oz + 3 oz = 17 oz

4. Add everything to get the total:

   304 oz + 17 oz = 321 oz

1. Add everything to get a total:

   3.43 kg + 1.90 kg + 3.57 kg = 8.9 kg

In this example, there are four operations in the NAS to the one operation in the metric system.

The advantage of using metric units to find out how much a container will hold can be even greater. An example of this is shown on the next page. (The most commonly used metric unit of volume is the liter— pronounced LEET-er and abbreviated l. *)

---

* We are underlining the abbreviation for liter (l) here to prevent possible confusion with the number one (1).

North American System

Metric System with SI Units

Problem: You have three contain-ers of apple cider:
    (a) 2 gallons 4 oz
    (b) 3 quarts 7 oz
    (c) 2 pints 3 oz
How much apple cider do you have?

Problem: You have three containers of apple cider:
    (a) 7.69 liters
    (b) 3.05 liters
    (c) 1.21 liters
How many liters of apple cider do you have?

1. Convert gallons to ounces:

    2 × 128 oz = 256 oz

1. Add to get the total:

    7.69 liters + 3.05 liters + 1.21 liters = 11.95 liters

2. Convert quarts to ounces:

    3 × 32 oz = 96 oz

3. Convert pints to ounces:

    2 × 16 oz = 32 oz

4. Add to get the total:

    256 oz + 96 oz + 32 oz + 4 oz + 7 oz + 3 oz = 398 oz

And so it goes. You probably agree that our familiar system of meas-uring things is really shamelessly—and needlessly—complicated.

Why is the metric system so much simpler? Because it uses the decimal system—because it is based on multiples of ten. If you're afraid you don't understand the decimal system, be assured that you do. You make use of it every day when you make change with money, since our money system is decimal. For example, a dime is one-tenth of a dollar, and a penny is one-hundredth of a dollar. So, one dollar plus one dime plus one penny can be written $1.11. If we want to buy eight things that cost $1.11 each, all we do is multiply $1.11 by 8:

    $1.11 × 8 = $8.88

But suppose our money system were based on a different system of units. Just for the sake of illustration, suppose there were 16 dimes (d) in a dollar or 120 pennies (¢) in a dollar. In this case, how would we write one dollar plus one dime plus one penny? How would we figure the price of eight things that cost this much ($1, 1 d, and 1¢)? First we would have to convert everything to a common unit:

$$\$1,\ 1\ d,\ 1\text{¢} = \$1 + \$\frac{1}{16} + \$\frac{1}{120}$$

Then we would have to perform two divisions:

$$\$\frac{1}{16} = \$0.0625 \qquad \$\frac{1}{120} = \$0.008333$$

Next we would have to add all three parts:

$$\$1 + \$0.0625 + \$0.008333 = \$1.071$$

Finally, we would have to multiply by 8 to get the total cost of the eight items.

$$\$1.071 \times 8 = \$8.568$$

No, it wouldn't be easy.* Fortunately, we don't need to go through these complications with our money since it is based on the decimal system. But there is no reason why we should do such complicated arithmetic in our other calculations either.

METRIC UNITS

All physical quantities have a number of common properties that can be measured. For example, they have matter (mass), they extend in space (length), and they endure in time. They also contain energy, maintain a temperature, emit light, and have electrical charge. Since these properties have always been there, you can well imagine the scores of different measuring systems we've had and also the measuring units. Length, for example, has been measured as hands, rods, leagues, yards, links, chains, inches, miles, paces, perches, furlongs, fathoms, fingers, and feet—to name only a few.

The only two important conditions for a measuring system are that the units be convenient and that they be consistent. Of all the measuring systems, metric units best satisfy these two conditions. But which metric units?

A list of all the units adopted as SI metric units appears in the Appendix. You needn't learn them all since many are used only in highly technical fields. Nor do you need to learn the basic metric unit of time because it is the second—just as in the NAS.

*Until recently, the people of Great Britain had to perform these mental acrobatics with their money.

1.   The most commonly used metric units,* and the ones you will need
     to learn, are:

> meter, for length;
> gram, for weight and mass;
> liter, for volume; and
> the degree Celsius, for temperature.

Study the table below and then do the exercise that follows.

Physical Quantity	Unit	SI Symbol
length	meter	m
weight and mass	gram	g
volume	liter	l
temperature	degree Celsius	°C

Without looking back, write in the name of the metric unit and the
symbol for each of these quantities:

Physical Quantity	Unit	SI Symbol
volume	_____	___
weight	_____	___
temperature	_____	___
length	_____	___

You can check your answers in the completed table above. Before
you continue to the next exercise, be certain that you have learned
these metric units and their symbols.

2.   The spellings we have used are American spellings. The British,
     naturally, spell things differently. (Why not? They invented the
     language.) These British spellings are given on the next page:

---

* There are some differences between the SI units basic for science
and the units used for everyday measurement. However, the everyday
units that we use here are consistent with the SI system. Also, the distinction between mass and weight is discussed later in the book.

U.S. Spelling	British Spelling
meter	metre
liter	litre
gram	gramme

So, if you are Canadian or work for certain American companies that prefer British spellings, you will spell "kilogram" as "kilogramme" and "centimeter" as "centimetre." It doesn't make any difference so long as you are able to recognize the variations. The conversion tables in the Appendix use British spellings simply because we reproduced them from an Australian book.*

The metric unit that is equal to 1000 kilograms is called a metric tonne and is written that way in England. What do you suppose is the American spelling of this unit? _____

– – – – – – – – – – – – – – – –

metric ton

METRIC PREFIXES

We have different units (like miles, yards, feet, and inches) because we need different-sized rulers, depending on whether we are measuring short, long, or very long distances. The same is true for other physical quantities. We could weigh everything in tons—if we wanted to. But when we wanted to buy a pound of steak, we would have to ask for 0.0005 ton. For convenience, we use different units to express different sizes. The only problem is that the units we have chosen to represent these sizes require difficult conversions. For example, we must learn that

$$12 \text{ inches} = 1 \text{ foot}$$
$$3 \text{ feet} = 1 \text{ yard}$$
$$5280 \text{ feet} = 1 \text{ mile}$$

And when we go to the supermarket, we must remember an entirely different set of conversions:

$$8 \text{ ounces} = 1 \text{ cup}$$
$$16 \text{ ounces} = 1 \text{ pint}$$
$$2 \text{ pints} = 1 \text{ quart}$$
$$4 \text{ quarts} = 1 \text{ gallon}$$

Each measure has its conversion—quarts, and bushels, and many other physical quantities, too. In practice, we remember the common

* B. Chiswell and E. C. M. Grigg, SI Units (Sydney, Australia: John Wiley & Sons Australasia Pty Ltd, 1971).

conversions. But for the entire span of our lives we must consult a guide for the measurements we don't use every day. This makes no sense at all.

With the metric system of prefixes and roots, the only conversions we need to make are from one size to another. There is a root for each physical quantity and a prefix for each significant size. To simplify things further, each prefix is a multiple of 10. For example, one common prefix is "kilo," meaning "a thousand times." This prefix added to the root "meter" (for the physical quantity of distance) produces "kilometer." A kilometer is a measurement of distance equal to 1000 meters. The prefix "kilo" added to the root "gram" (for the physical quantity of weight) produces "kilogram"—which is equal to 1000 grams.

The table below gives the most common prefixes and a memory aid (for most of them) to help you associate the prefix with the size. Note, too, that most of the prefixes that make the root smaller end in i. The prefixes that make the root larger end in o or a. The only prefix that rates a capital letter is the biggest one—Mega.

Common Metric Prefixes

Prefix	Memory Aid	Meaning
Makes Smaller		
deci– (DESS–ie)	The decimal system is in multiples of 10.	one tenth of (0.1)
centi– (SEN–ta)	The cent is one–hundredth of a dollar.	one hundredth of (0.01)
milli– (MILL–ie)	The millipede is a bug with 1000 legs.	one thousandth of (0.001)
micro– (MY–crow)	A microscope is used to see very small things.	one millionth of (0.000001)
Makes Larger		
hecto– (HECK–toe)	hecto – hundred	a hundred times (100)
kilo– (KILL–a)		a thousand times (1000)
Mega (MEG–a)	Megaphones increase the sound.	a million times (1,000,000)

The only prefixes you need to memorize are kilo-, centi-, and milli-.
It will be enough if you familiarize yourself with the others. The exercises that follow will help you.

3.   See if you can write the name of the prefix that is used to express each of the following sizes. Try not to look back.

                                                    Memory Aids

(a) 0. 000001 = ___Micro___   (An instrument used to measure small things)

(b) 100 times = ___Hecto___   (h)

(c) 0. 001 = ___Milli___   (a bug)

(d) 0. 1 = ___deci___   (decimal)

(e) 1000 times = ___Kilo___

(f) 0. 01 = ___centi___   (¢)

(g) 1, 000, 000 times = ___Mega___   (makes things loud)

— — — — — — — — — — — — — — — —

(a) micro- = 0.000001; (b) hecto- = 100 times; (c) milli- = 0.001;
(d) deci- = 0.1; (e) kilo- = 1000 times; (f) centi- = 0.01; (g) Mega-
= 1, 000, 000 times

4.   Change each of the following quantities to its equivalent in the root size. Don't look back. You can follow this example:

    1 decimeter = 0.1 meter

(a) 1 hectogram = ___100 GRAMS___

(b) 1 milliliter = ___.001 Liter___

(c) 1 kilometer = ___1000 meters___

(d) 1 Megagram = ___1,000,000 Grams___ (Don't forget that Mega- has a capital M.)

(e) 1 centiliter = ___.01 liters___

(f) 1 micrometer = ___.000001 meters___

— — — — — — — — — — — — — — — —

(a) 100 grams; (b) 0.001 liter; (c) 1000 meters; (d) 1, 000, 000
grams; (e) 0.01 liter; (f) 0.000001 meter

5. Unless you're sure you've got it, try again. Write the prefix names.

(a) 0.01 = _Centi_

(b) 0.001 = _Milli_

(c) 1000 times = _Kilo_

(d) 0.000001 = _Micro_

(e) 0.1 = _deci_

(f) 100 times = _Hecto_

(g) 1,000,000 times = _Mega_

– – – – – – – – – – – – – – –

(a) centi-; (b) milli-; (c) kilo-; (d) micro-; (e) deci-; (f) hecto-;
(g) Mega-

6. Just one more time. Change each quantity to its equivalent in the
root size. (Remember, you __must__ know the underlined prefixes.)

(a) 1 <u>milli</u>gram = _.001 Gram_

(b) 1 <u>centi</u>meter = _.01 Meter_

(c) 1 Megaliter = _1,000,000 Liters_

(d) 1 microgram = _.000001 Gram_

(e) 1 hectogram = _100 grams_

(f) 1 deciliter = _.1_

(g) 1 <u>kilo</u>meter = _1000 Meters_

– – – – – – – – – – – – – – –

(a) 0.001 gram; (b) 0.01 meter; (c) 1,000,000 liters; (d) 0.000001
gram; (e) 100 grams; (f) 0.1 liter; (g) 1000 meters

ABBREVIATIONS

The abbreviations, or symbols, for the metric prefixes and units are
handy to know. They're very simple, as the table on the next page shows.
Study all of them, and be sure to learn the abbreviation for kilo-, centi-,
and milli-. No doubt you will note that each symbol is the first letter
of the prefix or unit—except micro-. The symbol for micro- is μ, which
is the Greek letter <u>m</u>. (This was a necessary difference, given all the
units and prefixes that start with <u>m</u>.)

Metric Abbreviations and Symbols

Prefix	Symbol	Unit	Symbol
Mega-	M	meter	m
kilo-	k	gram	g
hecto-	h	liter	l
deci-	d		
centi-	c		
milli-	m		
micro-	μ		

7. You should be able to complete the following list. (The first two have been done for you.)

(a) 11 hectograms = 11 hg

(b) 1 millimeter = 1 mm

(c) 6 decigrams = ___6 dg___

(d) 20 micrometers = ___20 μ M___

(e) 900 kiloliters = ___900 Kl___

- - - - - - - - - - - - - - - -

(c) 6 dg; (d) 20 μm; (e) 900 kl

8. Without looking back at the table, abbreviate these units:

(a) deciliter = ___dl___

(b) milliliter = ___ml___

(c) hectogram = ___hg___

(d) centimeter = ___cm___

(e) micrometer = ___μm___

(f) kilometer = ___km___

(g) Megagram = ___Mg___

(h) centiliter = ___cl___

- - - - - - - - - - - - - - - -

(a) dl; (b) ml; (c) hg; (d) cm; (e) μm; (f) km; (g) Mg; (h) cl

CHAPTER THREE

Distance and Speed

> The meter rules where feet once tread,
> Enthroned beside the liter.
> O strange, indeed, for rhyme has said
> The foot will size the meter.

When asked to draw a 3-inch line without using a ruler, the average person will draw anything from $1\frac{1}{2}$ inches to 6 inches. (If you doubt this, try the experiment yourself on a group of friends.) Unfortunately, we're no better at estimating larger distances. If we could estimate distances with even fair accuracy, we could

- estimate walking, driving, or flying time;
- select the right screw for a home repair job;
- decide whether a table in the furniture store will be in the right proportion for a room;
- buy the right amount of cloth to make a skirt or cover a pillow;
- buy enough tile for a floor or enough paint for a wall;
- pick the right battery for a home appliance or a toy;
- determine whether a piece of furniture will fit through a door.

Once the metric system is fully adopted in North America, all cloth will be sold by the meter and the centimeter; speed limits will be designated by kilometers per hour (kph); screws, ammunition, and tools will be sized by the millimeter; and bathroom tiles will be measured by the centimeter. When you finish this chapter, you will be on intimate terms with these units. Equally important, you will have a much better idea of the size of a 15-millimeter hole, for example, than you now have of a 15-foot hole.

SMALL DISTANCES: MOST COMMON UNITS

Let's consider the small distances first—those we're most accustomed to measuring. In the metric system, the distances between a tenth of an inch and a few yards are normally expressed in three ways:

- in millimeters (mm);
- in centimeters (cm); or
- as decimal parts of a meter (m).

In scientific and engineering work, the millimeter and the meter are used frequently but the centimeter is seldom used. In everyday affairs, though, the centimeter is the most common unit. Here you will learn to use all three units interchangeably. Remember, a millimeter is equal to a tenth of a centimeter and a thousandth of a meter. As a handy reference, keep in mind that a meter is about 10 percent larger than a yardstick.

Here is a centimeter ruler:

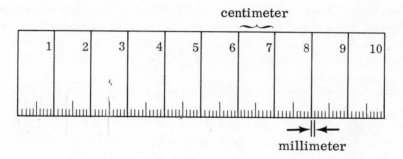

Each small division is 1 millimeter, and each large division is 10 millimeters or 1 centimeter.

1. You're going to make some measurements now to get a better idea of the unit sizes. It would be convenient if you had a centimeter ruler or a meter stick. (These are available in any stationery store or book store.) We have also made the outside edge of page 143 of this book a centimeter-millimeter ruler.

   (a) Measure the width of your pen or pencil. It is __6__ mm, __.6__ cm, or __.006__ m.

   (b) Measure the width of your thumb at the knuckle. It is __15__ mm, __.15__ cm, or __.015__ m.

   (c) Measure the length of your hand. It is __170__ mm, __17__ cm, or __.17__ m.

(d) Remember, there are <u>fewer</u> meters in a football field than there are yards. So, the length of a football field (100 yards long) is

roughly _____ mm, _____ cm, or _____ m.

- - - - - - - - - - - - - - - -

Check your answers. If they compare favorably with the suggestions below, you may go ahead to the next exercise. If not, perhaps you should review Chapter 2 before you go on.

(a) 5-8 mm, 0.5-0.8 cm, 0.005 -0.008 m
(b) 15-35 mm, 0.15-0.35 cm, 0.015-0.035 m
(c) 150-250 mm, 15-25 cm, 0.15-0.25 m
(d) 90,000 mm, 9000 cm, 90 m

LARGE DISTANCES: REFERENCES FOR THE METER

Most of the world's measurement systems came from common, practical reference sizes. The foot obviously refers to the size of a human foot. The <u>Bible</u> mentions the cubit (the length of the forearm, or about 18 to 20 inches), which was a common ancient measure. For example, Goliath was 6 cubits tall (9 feet), and Noah's ark is described as 300 cubits long (600 feet, or about half the length of the <u>Queen</u> <u>Elizabeth</u> <u>II</u>).

Metric lengths were established more scientifically. Initially, in the eighteenth century, the meter was defined as 1/10,000,000 of the distance between the Equator and the North Pole. Since then, a more precise standard has been established. Nevertheless, we can still say there are about 10,000,000 meters from the Equator to the North Pole.

2. Use this reference now to complete the following statements:

(a) There are _____ kilometers from the Equator to the South Pole.

(b) The distance around the world at the Poles would be very

nearly _____ km.

- - - - - - - - - - - - - - -

(a) 10,000 km; (b) 40,000 km

2,54 cm
/ in

SMALL DISTANCES: BODY REFERENCES

Our bodies give us easy references for small distances. The writers
(one a woman, one a man) found that when each pressed four fingers
together to form a straight edge, the measure was 50 mm:

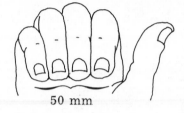

50 mm

Press your fingers together like this and measure them. If your fingers
are particularly large, you may have to press them closer together or
use three fingers to measure 50 mm. If they are particularly small, you
may have to spread them slightly or use the thumb, too. Once you see
how to hold your fingers to measure 50 mm, you will have a ruler that
will always be with you.

Study the Table of Body Reference Sizes. Measure yourself and
make any necessary adjustments. Then memorize the different lengths.

Table of Body Reference Sizes

Size	Body Reference	
10 mm		Width of the tip of the little finger, or the width of a fingernail
25 mm		Tips of two fingers pressed together
50 mm		Tips of four fingers pressed together

Size	Body Reference
100 mm	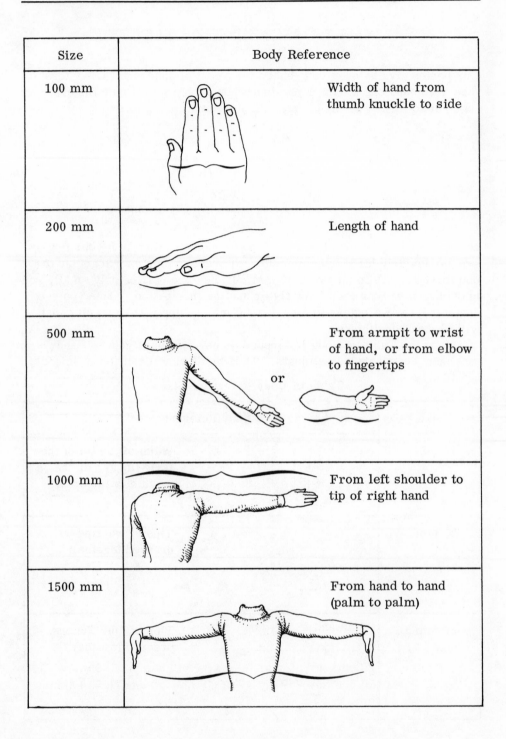 Width of hand from thumb knuckle to side
200 mm	Length of hand
500 mm	From armpit to wrist of hand, or from elbow to fingertips or
1000 mm	From left shoulder to tip of right hand
1500 mm	From hand to hand (palm to palm)

3. Write the size in millimeters, centimeters, or meters—whatever unit is asked for—next to each body reference.

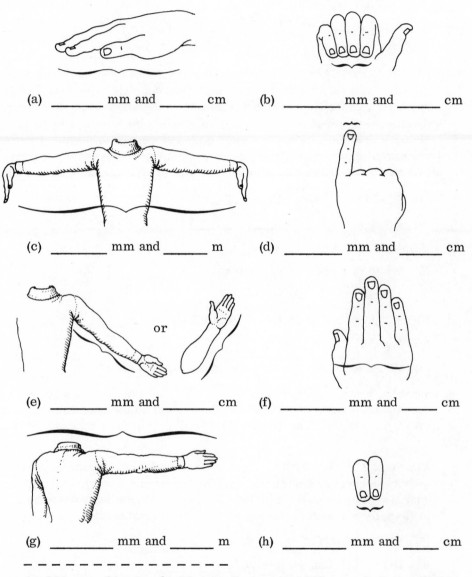

(a) _____ mm and _____ cm   (b) _____ mm and _____ cm

(c) _____ mm and _____ m   (d) _____ mm and _____ cm

(e) _____ mm and _____ cm   (f) _____ mm and _____ cm

or

(g) _____ mm and _____ m   (h) _____ mm and _____ cm

_ _ _ _ _ _ _ _ _ _ _ _ _ _ _ _ _ _

(a) 200 mm, 20 cm; (b) 50 mm, 5 cm; (c) 1500 mm, 1.5 m; (d) 10 mm, 1 cm; (e) 500 mm, 50 cm; (f) 100 mm, 10 cm; (g) 1000 mm, 1 m; (h) 25 mm, 2.5 cm

4. If you are having trouble memorizing any of these body references, here is some help. Start with your little finger (10 mm), and count off as you move along your body:

ten,   twenty-five,   fifty, and so on.

Once you have your own personal reference sizes, practice using them. We'll do that now. Estimate the sizes of the following distances. Be careful to distinguish between millimeters (mm) and centimeters (cm). Then use your body references to test your estimates. Finally, actually measure to check your answers. (We've given some approximate answers for you to compare with yours.)

(a) Your shoe length _____ mm

(b) From knee to floor _____ cm

(c) Length of a closed mouth _____ mm

(d) Width of a package of cigarettes _____ cm

(e) Length of a telephone receiver _____ mm

(f) Height of the chair from seat to floor _____ mm

(g) Length of your pencil _____ cm

(h) Width (diameter) of the eraser on your pencil _____ mm

- - - - - - - - - - - - - - - -

Approximate answers: (a) 250-350 mm; (b) 50 cm; (c) 60 mm; (d) 5 cm; (e) 200 mm; (f) 400 mm; (g) 19 cm; (h) 7 mm

5. Try some others. Again, try to estimate without thinking of your body references. Then write your answers and check them with your body references. Finally, compare them with our answers or actually measure. (Be sure to watch your decimals!)

(a) Width (diameter) of a quarter _____ cm

(b) Diameter of an aspirin tablet _____ mm

(c) Length of a king-size cigarette _____ mm

(d) Diameter of an LP record _____ cm

(e) Length of a house key _____ cm

(f) Diameter of a hole in the telephone dial _____ mm

(g) Width of your kitchen sink _____ mm

(h) Width of the nearest window _____ cm

(i) Length of a dinner fork _____ cm

(j) Height of a table _____ mm

(k) Length of a city telephone directory _____ mm

(l) Diameter of a cigarette _____ mm

(m) Length of the room you are in _____ cm

(n) Length of your desk or table _____ cm

(o) Height of a can of cola _____ cm

- - - - - - - - - - - - - - -

Approximate answers for those items that are constant: (a) 2.5 cm;
(b) 10 mm; (c) 80 mm; (d) 30 cm; (e) 5 cm; (f) 13 mm; (i) 19 cm;
(k) 280 mm; (l) 8 mm; (o) 12 cm

SMALL DISTANCES: OPTIONAL EXERCISES

6. This exercise can help you to become very accurate in estimating
lengths under 300 millimeters. You'll need only your millimeter
ruler and a large sheet of paper. Follow these instructions very
carefully:

(a) Without looking at the ruler, draw a line that you think is equal
to the line specified under "Length" in the score sheet on page
26. Start with Group A, Trial 1. Use your body reference to
check your drawing.

(b) Measure your line with a ruler and enter this length under the
column "Your Score" on the score sheet. Follow the same pro-
cedure for the other four trials of Group A.

(c) At the end of the five trials of Group A, add up your scores and
subtract your total from the correct total given on the score
sheet. Enter this difference on the score sheet.

(d) Repeat this process for additional groups of five trials each
until you are satisfied with your accuracy.

Remember, don't look at the ruler when you draw the lines. And
try not to use your body references when you are estimating. Use
them only to check your work before you actually measure with the
ruler. Now begin by drawing a line equal to 1 mm—the length specified for Trial 1, Group A. Space is provided on the next page.

SCORE SHEET

Trial	Length (mm)	Your Score	Trial	Length (mm)	Your Score
(A) 1	1		(B) 1	213	
2	130		2	84	
3	85		3	151	
4	7		4	3	
5	260		5	112	
Total	483		Total	563	
Difference			Difference		
(C) 1	96		(D) 1	120	
2	13		2	235	
3	176		3	35	
4	34		4	166	
5	3		5	8	
Total	322		Total	564	
Difference			Difference		
(E) 1	62		(F) 1	147	
2	181		2	9	
3	4		3	190	
4	221		4	51	
5	106		5	73	
Total	574		Total	470	
Difference			Difference		

(G)	1	18		(H)	1	33	
	2	47			2	2	
	3	91			3	172	
	4	206			4	106	
	5	38			5	58	
	Total	400			Total	371	
	Difference				Difference		
(I)	1	31		(J)	1	124	
	2	17			2	108	
	3	190			3	87	
	4	79			4	16	
	5	113			5	39	
	Total	430			Total	374	
	Difference				Difference		

7. Here you can practice estimating distances from 200 mm to 1500 mm. For this exercise, you will need a long piece of string 2000 mm or longer. Follow these instructions:

(a) Sample lengths are on the score sheet on the next page. Estimate the length given for Trial 1, Group A, on the piece of string; put a knot in it. You can use your body references to check your work.

(b) Measure your length of string with a ruler and enter this length under "Your Score." Do the same for the other four trials of Group A.

(c) At the end of the five trials of Group A, add up your scores and subtract your total from the one given. Enter this difference on the score sheet.

(d) Repeat this process for additional groups of five trials each until you find you are fairly accurate.

Begin now by estimating 325 mm on the piece of string.

SCORE SHEET

Trial	Length (mm)	Your Score	Trial	Length (mm)	Your Score
(A) 1	325		(B) 1	980	
2	215		2	440	
3	1100		3	210	
4	625		4	1100	
5	450		5	355	
Total	2715		Total	3085	
Difference			Difference		
(C) 1	312		(D) 1	660	
2	1000		2	555	
3	400		3	340	
4	720		4	275	
5	645		5	875	
Total	3077		Total	2705	
Difference			Difference		
(E) 1	1200		(F) 1	675	
2	915		2	1115	
3	390		3	660	
4	860		4	345	
5	480		5	500	
Total	3845		Total	3295	
Difference			Difference		

(G)	1	275		(H)	1	420	
	2	775			2	585	
	3	1100			3	210	
	4	620			4	1225	
	5	410			5	680	
	Total	3180			Total	3120	
	Difference				Difference		
(I)	1	660		(J)	1	775	
	2	350			2	650	
	3	990			3	1000	
	4	225			4	430	
	5	165			5	335	
	Total	2390			Total	3190	
	Difference				Difference		

CIRCUMFERENCE

8. Do you want to buy a wood screw, a curtain rod, or a drill bit?
    How about an electrical wire, a ring, or a belt? Many of these are
    already manufactured in metric units.

    Circumferences are particularly difficult to estimate because
    we are so used to thinking of distance as a straight line. The task
    will be simpler if we recall the formula for circumference:

    Circumference = π (pi) × Diameter

    Remember, π is approximately 3 1/7, or 3.141592. The formula
    means that the distance around any circle is a little more than three
    times the distance across it.

    As an example, an aspirin tablet is about 10 mm across.* The
    same aspirin will be a bit more than 30 mm in circumference, or
    31.41592 mm.

    Estimate the circumferences of the items on the next page:

---

* Some manufacturers, like Bayer's, make their aspirins exactly 1 cm
in diameter—not a bad reference unit if memorizing gives you a headache.

		Across	Approximate Circumference	
(a)	Pencil	8 mm	_____	mm
(b)	Needle	1 mm	_____	mm
(c)	A human head	160 mm	_____	cm
(d)	A can of cola	67 mm	_____	mm
(e)	Tree	1 m	_____	cm
(f)	Thread	0.3 mm	_____	mm
(g)	Aspirin tablet	1 cm	_____	cm

– – – – – – – – – – – – – – – –

Approximate answers: (a) 26 mm; (b) 3.14 mm; (c) 50 cm; (d) 210 mm; (e) 314 cm; (f) 1 mm; (g) 3.14 cm

9.   You can now see that if you can estimate lengths, you can easily guess the circumference of a circle. Try your skills on these familiar objects. (And watch your mm's and cm's!)

		Across		Circumference	
(a)	Head of a pin	_____	mm	_____	mm
(b)	Hole in a door key	_____	mm	_____	mm
(c)	Hole in the telephone dial	_____	mm	_____	cm
(d)	A quarter	_____	mm	_____	mm
(e)	A nickel	_____	mm	_____	mm
(f)	A cigarette	_____	mm	_____	cm
(g)	Your thigh	_____	mm	_____	cm
(h)	Your chest	_____	cm	_____	cm
(i)	Your thumb	_____	mm	_____	cm
(j)	Your wrist	_____	cm	_____	mm
(k)	Concentrated soup (ordinary-sized can)	_____	mm	_____	cm
(l)	Your little finger	_____	mm	_____	cm

– – – – – – – – – – – – – – –

Approximate answers: (a) 1.5-5 mm; (b) 5-15 mm; (c) 10 mm - 3 cm; (d) 25-75 mm; (e) 20-60 mm; (f) 7 mm - 2.2 cm; (g) 120-200 mm, 37-60 cm; (h) 25-40 cm, 75-120 cm; (i) 20 mm - 6 cm; (j) 5-8 cm, 150-250 mm; (k) 8-25 cm; (l) 8-10 mm, 2.5-5 cm

10. This optional exercise will provide extra practice in estimating circumference. It will be particularly useful if you're in a machine trade.

On a separate sheet of paper, draw circles with the circumferences given on the score sheet below. Then actually measure the diameters of these circles and multiply by three to get your scores. Finally, add the subtotals for the trials in each group, and subtract your total from the correct total. Continue until you are satisfied with your accuracy.

SCORE SHEET

Trial	Circum-ference	Your Score	Trial	Circum-ference	Your Score
(A) 1	1 mm		(B) 1	25 mm	
2	3 cm		2	10 mm	
3	10 cm		3	55 mm	
4	6 mm		4	20 mm	
5	240 mm		5	5 cm	
Total	377 mm		Total	160 mm	
Difference			Difference		
(C) 1	20 mm		(D) 1	35 mm	
2	10 cm		2	16 cm	
3	7 cm		3	140 mm	
4	58 mm		4	60 mm	
5	24 cm		5	70 mm	
Total	488 mm		Total	465 mm	
Difference			Difference		

(E)	1	60 mm		(F)	1	34 mm	
	2	35 mm			2	24 mm	
	3	12 cm			3	10 cm	
	4	62 mm			4	42 mm	
	5	15 cm			5	80 mm	
	Total	427 mm			Total	280 mm	
	Difference				Difference		

INTERMEDIATE DISTANCES: HEIGHT AS A REFERENCE

Now that you can think in millimeters and centimeters, let's turn to meters. Recall your body reference for a meter—left shoulder to the tips of the right fingers. Another useful body reference is your height. Below is a table of heights. Find your height in centimeters (rounded off to the half-centimeter).

Table of Heights

North American System		Metric	North American System		Metric
ft	in.	cm	ft	in.	cm
4	9	145	5	9	175.5
4	10	147.5	5	10	178
4	11	150	5	11	180.5
5	0	152.5	6	0	183
5	1	155	6	1	185.5
5	2	157.5	6	2	188
5	3	160	6	3	190.5
5	4	162.5	6	4	193
5	5	165	6	5	195.5
5	6	167.5	6	6	198
5	7	170	6	7	200.5
5	8	172.5	6	8	203

Memorize your own metric height. Try to remember how tall a few relatives and friends are—and also some children. These heights will be handy references, too. An additional set of references might be the so-called average heights for men and women. For women the average height is 165 cm (5 ft 4 in.), and for men it is 180 cm (5 ft 10 in.).

11. Without looking back at the Table of Heights, complete the following sentences:

(a) Your height in millimeters is _____ mm.

(b) The average woman is about how tall in centimeters?

_____ cm

(c) The average man is about _____ cm.

(d) A basketball player would most likely be about _____ cm.

(e) A jockey would be about _____ cm.

- - - - - - - - - - - - - - - -

Approximate answers: (a) your own answer; (b) 165 cm; (c) 180 cm; (d) 200 cm; (e) 150 cm

INTERMEDIATE DISTANCES: A PACE AS A REFERENCE

A pace is another convenient way to estimate distances in meters. With very little practice, you can estimate the length of a room to within a few centimeters. Take the time to measure the dimensions of the room you're in right now, and then practice pacing it off until you get the feel of a 1-meter stride.

$1 m = 100 cm = 1000 mm$

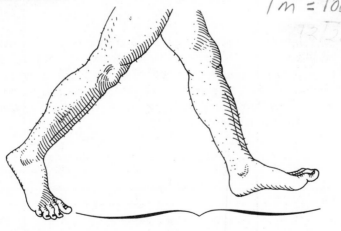

1 meter

Remember your body reference for a meter and use it to determine how long your meter stride should be. If you are used to pacing off yards, add about 10 percent to your stride, since a meter is 10 percent longer than a yard. The pace will become a handy, lifetime meter ruler. (But, caution! Forget about yards. Think metric!)

12. In this exercise you are not expected to get answers that agree with ours exactly. You will get some practice in visualizing intermediate meter distances. Make the best estimates you can by pacing off the distance whenever this will help. If you wish, you can actually measure to check your estimates.

    (a) Length of your bedroom, or the room you are in _____ m

    (b) Width of your bedroom, or the room you are in _____ m

    (c) Height of a kitchen table _____ m

    (d) Height of a stove _____ m

    (e) Height of a door _____ m

    (f) Width of the nearest window _____ m

    (g) Height of the nearest window _____ m

- - - - - - - - - - - - - - - - -

The following are our answers. Unless your surroundings are most unusual, your answers should be close to ours.

(a) 5.5 m; (b) 4.25 m; (c) 0.9 m; (d) 0.75 m; (e) 2 m; (f) 0.9 m; (g) 1.9 m

INTERMEDIATE DISTANCES: OTHER REFERENCES

One interesting landmark you can choose to remember is the height of the tallest building in your town, or a town near you. International travelers might choose to remember the height of the Eiffel Tower, or the Tower of London, or the Taj Mahal. The table below and on the following pages shows the heights of some buildings in the U.S. and in Canada. Study this table for a few minutes.

Height of Buildings in U.S. and Canada

City	Building	Height (m)
Atlanta	Peachtree Center Hotel	212
Austin	Capitol	94

Baltimore	U.S. Fidelity	161
Birmingham	1st National Southern	95
Boston	John Hancock	241
Buffalo	Marine Midland	161
Calgary	Husky Tower	191
Charlotte	Jefferson 1st Union	132
Chicago	Sears Tower	442
Cincinnatti	Carew Tower	175
Cleveland	Terminal Tower	216
Columbus	30 E. Broad St.	189
Dallas	1st International	216
Dayton	Winters Bank	123
Denver	Brooks Towers	128
Des Moines	Equitable	97
Detroit	Penobscot	170
Ft. Wayne	Nat'l Bank	103
Ft. Worth	Nat'l Bank	138
Halifax	Fenwick Towers	91
Hartford	Travelers Ins.	161
Honolulu	Ala Moana Hotel	119
Indianapolis	Indiana National Bank	218
Jacksonville	Independent Life	154
Kansas City	Light & Power	145
Las Vegas	International Hotel	105
Los Angeles	United City Bank	256
Louisville	1st Nat'l Bank	156
Memphis	100 N. Main	131
Miami	One Biscayne Corp	139
Milwaukee	1st Wisconsin	183
Minneapolis	IDS Center	235

Montreal	Place Victoria	190
Nashville	Nat'l Life & Accident	138
New Orleans	1 Shell Square	212
New York	World Trade Center	412
New York	Empire State	381
Oakland	Ordway	123
Oklahoma City	Liberty Tower	152
Omaha	Woodmen Tower	138
Ottawa	Parliament	92
Philadelphia	City Hall	167
Phoenix	Valley Nat'l Bank	147
Pittsburgh	U.S. Steel	256
Portland, Ore.	1st Nat'l Bank	164
Providence	Industrial Nat'l Bank	128
Richmond	City Hall	95
Rochester	Xerox Tower	135
St. Louis	Gateway Arch	192
St. Paul	1st Nat'l Bank	123
Salt Lake City	LDS Church Office Bldg	128
San Antonio	Tower of Americas	190
San Diego	So. Calif. 1st Nat'l Bank	118
San Francisco	Transamerica	260
Seattle	1st Nat'l Bank	186
Syracuse	State Tower	96
Tampa	1st Finance Tower	140
Toledo	Owens-Corning	122
Toronto	Canadian Imperial Bank	239
Tulsa	1st Nat'l Bank	157
Vancouver	British Columbia Center	208

Washington	Capitol	170
Winnipeg	Lombard Place	124
Winston–Salem	Wachovia	125

Another good reference is the city block, which is usually between 50 and 100 meters long (50 meters north to south in New York City, and 100 meters in the center of a planned city like Columbia, South Carolina).

If you are sports-minded, you know that many track and field measures are metric. An example is the 100-meter race, which is equivalent to the 100-yard dash, or the 1500-meter race, which is like the mile race. Here are some other references in sports:

		Round off to remember
Football field	91.5 m	90 m
Home plate to left-field wall	100 m	100 m
Pitcher's mound to home plate	18.44 m	18 m
Furlong	201.2 m	200 m
Olympic pole vault	563.9 cm	560 cm
Olympic long jump	838.2 cm	840 cm
Olympic high jump	222.9 cm	225 cm
Typical golf fairway	350 m	350 m
Height of a basketball player	200 cm	200 cm
Men's ski jump	90 m	90 m
Women's ski jump	70 m	70 m

SPEED AND GREAT DISTANCES

We translate great distances into time. How long does it take to get there? Here you will learn to think of distances as distances in time, specifically as kilometers per hour (km/h), the everyday measure. Most scientific work uses meters per second (m/s). The translation from kilometers per hour (km/h) to meters per second (m/s) is not easy to remember, since you must multiply km/h by 0.278 to get m/s. If you do need to make this translation, the conversion tables in the Appendix will be useful.

All road signs will one day give distances in kilometers and speed limits in kilometers per hour, as they do now in most other countries. If you drive a foreign car, you are probably already accustomed to speedometer readings in kilometers per hour.

Let's consider first how to get a good idea of metric speeds without making any conversions. Think of a car as having 100 km/h allowed before it breaks the speed limit. If you go half of that allowable speed, you are driving in the city (50 km/h). If you go one-quarter of the allowed speed (25 km/h), you are driving in a school zone.

The table below shows these and other references for speed. Study them carefully.

References for Speed

Speed (km/h)	Activity	% of Speed Limit
5	Walking at a moderate pace	5%
10	Jogging	10%
25	Driving in a school zone, first or second gear	25%
50	City driving, light traffic	50%
75	Driving on a 2-lane road, approaching the city	75%
100	Thruway driving	100%
125	Speeding on the thruway	125%
150	High-speed train	150%

Memory device

13. Without looking back at the table, see if you can match each activity
    with the most likely speed reference.

    _____ (a) High-speed Olympic running   1000 km/h

    _____ (b) Driving the Indianapolis Speedway   150 km/h

    _____ (c) Driving by horse and carriage   100 km/h
                    through Central Park in New
                    York City   250 km/h

    _____ (d) Traveling on a high-speed train   25 km/h

    _____ (e) Traveling on a jet plane   10 km/h

                                                          75 km/h

    - - - - - - - - - - - - - - - -

    (a) 25 km/h; (b) 250 km/h; (c) 10 km/h; (d) 150 km/h; (e) 1000
    km/h

14. Answer the following questions:

    (a) How long would it take to drive on the thruway to a city 350 km

         away? _____ hours

    (b) How long would it take to walk 10 km? _____ hours

    (c) How long does it take a high-speed train to travel 1500 km

         nonstop? _____ hours

    (d) How fast does a jogger go when he's jogging? _____ km/h

    (e) How fast do you drive in light city traffic? _____ km/h

    (f) How fast do you drive through a school zone? _____ km/h

    (g) How fast do you drive on the approaches to the city?

         _____ km/h

    (h) What is the normal thruway speed? _____ km/h

    - - - - - - - - - - - - - - - -

    (a) 3.5 hours; (b) 2 hours; (c) 10 hours; (d) 10 km/h; (e) 50 km/h;
    (f) 25 km/h; (g) 75 km/h; (h) 100 km/h

The chart below gives some approximate distances between major North American cities—given in kilometers, of course. They might be useful references to you.

	Atlanta	Boston	Chicago	Dallas	Los Angeles	Mexico City	Miami	Montreal	New Orleans	New York	Seattle	Washington
Atlanta		1718	1174	1295	3534	2896	1070	1979	793	1376	4434	1014
Boston	1718		1569	2927	4911	4669	2481	531	2471	348	4885	703
Chicago	1174	1569		1506	3371	3332	2188	1364	1519	1187	3319	1146
Dallas	1295	2927	1506		2257	1849	2106	2837	801	2576	3398	2208
Los Angeles	3534	4911	3371	2257		3245	4364	4698	3059	4690	1974	4254
Mexico City	2896	4669	3332	1849	3245		3543	4579	2164	4219	4743	3855
Miami	1070	2481	2188	2106	4364	3543		2743	1409	2140	5504	1778
Montreal	1979	531	1364	2837	4698	4579	2743		2639	624	4340	965
New Orleans	793	2471	1519	801	3049	2164	1409	2639		2132	4200	1768
New York	1376	348	1187	2576	4690	4219	2140	624	2132		4672	368
Seattle	4434	4885	3319	3398	1974	4743	5504	4340	4200	4672		4422
Washington	1014	703	1146	2208	4254	3855	1778	965	1768	368	4422	

LAND AREA

In metric countries, the basic unit of land area is the <u>are</u> (pronounced AIR), which is 100 square meters, or 10 meters on each side.

10 m

1 are = 100 m^2

(10 m on the left side)

An ordinary one-story suburban house would cover about 2 ares. However, metric real estate is sold in hectares (100 ares), equal to 10,000 square meters. (See the diagram on the next page.) One square kilometer equals 100 hectares. One hectare is 2.5 times the size of an acre. An ordinary suburban lot would be about one-fifth of a hectare, but it will be some time before the hectare becomes a common unit of measure in the United States. Think of all the property deeds that must be changed!

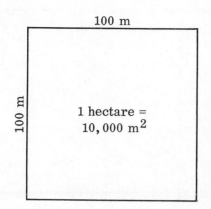

Still, the metric system will make people more aware of real estate values. For example, a familiar television commercial advertises "Four lots: only $5000 for all four." These lots, as it turns out, are 75 feet by 50 feet. All four together amount to only about one-third of an acre. This means that the property is actually being sold for $14,520 an acre—rather a hefty sum!

The problem is, of course, the difficulty of remembering conversion units in our North American System and then calculating the cost. The metric system will make it easier for us to spot deceptions. For example, in the metric system all we need to know is the value of a square meter. If it were $10, we would know immediately what a half-hectare would cost (5000 × $10 = $50,000).

15. You can complete this chapter with a summary exercise that provides additional practice in measuring shorter lengths. Using a separate sheet of paper, draw lines of the dimensions specified on the score sheet. Remember, add up your subtotals to get a total for each group of five trials, and compare this total with the correct one. The difference will show you how accurate you are. Continue until you are satisfied with your accuracy.

The score sheet is on the following page.

SCORE SHEET

	Trial	Length (mm)	Your Score		Trial	Length (mm)	Your Score
(A)	1	1		(B)	1	91	
	2	10			2	8	
	3	180			3	33	
	4	35			4	114	
	5	22			5	19	
	Total	248			Total	265	
	Difference				Difference		
(C)	1	4		(D)	1	86	
	2	225			2	148	
	3	62			3	12	
	4	30			4	106	
	5	167			5	42	
	Total	488			Total	394	
	Difference				Difference		

CHAPTER FOUR

Volume

. . . The loathesomeness of all mankind has become plain
to me, for through them the earth is full of violence. I in-
tend to destroy them and the earth with them. Make yourself an ark with ribs of cypress; cover it with reeds and
coat it inside out with pitch. This is to be its plan: the
length of the ark shall be three hundred cubits, its breadth
fifty cubits, and its height thirty cubits. You shall make a
roof for the ark, giving it a fall of one cubit when complete;
and put a door on the side of the ark, and build three decks,
upper, middle, and lower.

Genesis 6:4

God instructed Noah to build an ark to a size of 300 by 50 by 30 cubits,
and with three decks (first class, tourist, and steerage?). The ark, then,
was 450,000 cubic cubits, or about 50,000 cubic meters (m^3). If we
assume that "birds, beasts, and reptiles" average the size of a house
cat, then Noah would have had space for over 2 million species (in pairs).*
There are only about 20,000 species of reptiles, amphibians, birds, and
mammals. So Noah had more than enough space (though the smell during
those damp days must have discouraged much roaming about). So much
for the skeptics who doubted the ark's capacity. These skeptics clearly
are unable to speak the language of measurement. As this example
shows, volume is deceptive indeed.

Can you guess how many eggs a 1-cubic-meter carton will hold?
(The carton would come up to your navel if you are 165 cm tall, or up to

---

* Assume a cat is 0.4 m long, 0.1 m wide, and 0.25 m tall. He then
occupies 0.01 m^3 space (0.4 × 0.1 × 0.25). If you divide 50,000 m^3 by
0.01 m^3, you get 5 million m^3. Since Noah required two of each species,
he could have stored 2.5 million kinds of animals. Even if the average
animal were the size of a medium-sized dog, Noah could have stored
well over 100,000 species.

your hip bone if you are 180 cm tall.) This carton can hold over 10,000 eggs, although most people guess only 500 to 1000. The next time you go to the fair, guess 10,000 beans in the 2-liter bottle and you just might win the prize!

But some people are very good estimators of volume. For example, Grandma's recipes are legendary: "Take some flour, mix it with a little water, add some salt . . . ." Grandma has a sense of the right measures. She knows about how much flour is in a cup and about how much salt makes a teaspoonful. So do all creative cooks. Have you ever watched Julia Child add salt to a recipe?

Metrication (conversion to the metric system) will create a demand for new cookbooks and new cooking utensils, all in metric units. Soon we will be able to interpret metric recipes even more easily than we now interpret the recipes in the usual cookbooks. However, creative cooking —improvising and substituting—will be impossible unless we can develop a sense of metric measures.

We need to estimate volume or capacity in areas other than cooking. A good consumer probably knows roughly how much cola is in a 12-ounce can or how much food a 15.3-cubic-foot refrigerator will hold. This chapter will teach the consumer to make estimates like these in metric units.

UNITS OF VOLUME

Our measuring system has two separate sets of measures for volume, one dry and one liquid. Confusion is increased by units like the dry quart and the liquid quart, which are very nearly—but not exactly—the same. Metric measures for volume are much simpler because one set of units is used for both the liquid and the dry measures. The metric system uses three main units to measure volume:

- liter (l);
- milliliter (ml); and
- cubic meter (m^3).

The liter is most common, probably because it is so much like a quart. (It is actually 5 percent greater than a quart.) By definition, 1 liter is 1 cubic decimeter (1 dm^3).* In other words, a cube with each side equal to 1 decimeter (or 10 cm) will hold 1 liter. The diagram on the next page illustrates the liter measure.

Most familiar household items will be measured or sized in liters. Pots, for example, will come in liter sizes, half-liter sizes, or 2-liter

---

*It is very nearly 1 dm^3, but not quite; so the liter cannot be used in high-precision scientific work.

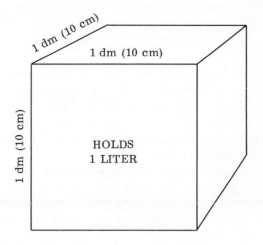

sizes. Milk will be in 1-liter, 2-liter, or 4-liter containers. Beer will probably be in half-liter cans or liter cans, and so will soda. Instead of the dry quarts and the dry bushels, farmers will package their strawberries in liter boxes and their potatoes in cubic-meter baskets.

The milliliter, reserved for much smaller measures of volume than the liter, is one-thousandth of a liter and is equal to 1 cm^3, or 1 cubic centimeter. In other words, a cube with each side equal to 1 cm will hold a quantity of 1 ml (or 1 cm^3). (Sometimes you will see cm^3 written as cc or ccm—for cubic centimeter.)

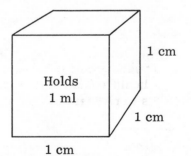

When metrication is complete, you will probably buy Worcestershire sauce and all antiseptics in milliliter bottles. As a matter of fact, most liquid medicines are already sold in milliliters.

The cubic meter, or m^3, is used for very big measures—from the volume of water in a home storage tank to the capacity of a railway freight car. So, a cube with each side equal to 1 m will store 1 m^3. A diagram illustrating this measure is on the following page.

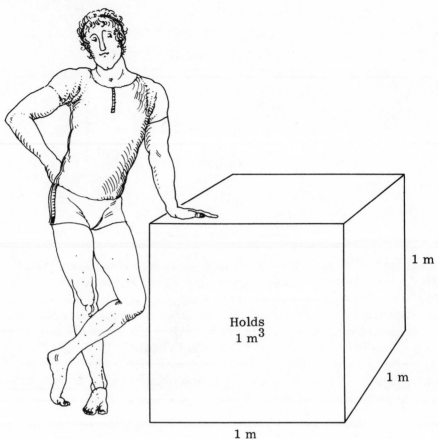

Holds
1 m^3

1 m

1 m

1 m

1.   Which unit (liter, milliliter, or cubic meter) will be most useful in each of the following situations?

(a)   Measuring the volume of medicine to give to a patient

_____

(b)   Packaging blackberries   _____

(c)   Determining the amount of space in a refrigerator (i.e., how much food it will store)   _____

– – – – – – – – – – – – – – – –

(a) milliliter;   (b) liter;   (c) cubic meter

ESTIMATING VOLUME

Remember, volume can be very deceptive. A pot with a volume (or capacity) of 1 liter doesn't look as if it will hold half as much as a pot with a volume of 2 liters:

These two pots show that small changes in the dimensions of a container will make very great differences in the amount of material that container will store.

A cube that is 4 meters on the edge is

$$4 \text{ m} \times 4 \text{ m} \times 4 \text{ m} = 64 \text{ m}^3$$

But a 5-meter cube is

$$5 \text{ m} \times 5 \text{ m} \times 5 \text{ m} = 125 \text{ m}^3$$

In this example, by increasing the length of a side by 20 percent, we almost double the volume.

THE LITER

You will probably agree that the way to develop a good sense of a metric unit is to associate it either with a part of the body or with a familiar object. The capacity of a carton of milk is something familiar that you can associate with a liter (or 1000 milliliters). (Actually, a liter is 5 percent larger than a quart.) The carton can be divided as in the diagram on the next page:

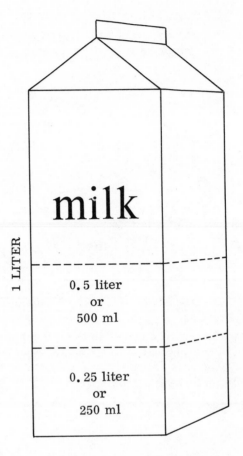

2. You should be able to answer the following questions:

(a) About how much milk will a gallon jug hold? *4 liters*

(b) How much milk is in a pint-sized carton? *500 ml 0.5 ml* *2 cups or*

(c) How many raspberries will a pint-sized basket hold?

_____

- - - - - - - - - - - - - - - - -

(a) 4 liters; (b) 500 milliliters or 0.5 liter; (c) 500 milliliters or 0.5 liter

3. Try to complete these statements without looking back:

(a) A glass of milk, or one-fourth of a carton, holds about _____

_____ liter(s).

(b)   Half a carton of milk holds about _____ liter(s).

(c)   A volume of 1000 milliliters is equivalent to __*l*__ liter(s).

(d)   A full carton of milk, or a quart, holds about __*l*____
      liter(s).

(e)   A volume of 1 liter is equivalent to _*1000*_ milliliter(s).

(f)   A volume of 0.5 liter is equivalent to _*500*_ milliliter(s).

(a) 0.25 liter; (b) 0.5 liter; (c) 1 liter; (d) 1 liter; (e) 1000 milliliters; (f) 500 milliliters

OTHER MEASURES OF VOLUME

A milk carton and its divisions should help you with the liter, which is
the most common measure of volume. But what about the small volumes
and the very large ones?

First, let's consider the small volumes, measured in milliliters.
A good way to remember what a milliliter (a cubic centimeter) looks like
is to associate it with the tip of your little finger—the "pinky." Remember, the width of your smallest finger or fingernail is roughly 1 cm.
The length of the fingernail is about as long, so the tip of the finger
itself down to the beginning of the fingernail is a box that holds roughly
1 cm × 1 cm × 1 cm, or 1 cm^3 (also written 1 cc or 1 ccm).

Two other small-volume references are particularly useful:

$$1 \text{ teaspoonful} = 5 \text{ ml}$$
$$1 \text{ tablespoonful} = 15 \text{ ml}$$

Large volumes are measured in cubic meters. In Chapter 3 we said
that the meter is the distance from your left shoulder to your right finger-
tip, with your arm outstretched. A cubic meter (m^3), then, is the
amount of space taken up by a box 1 m long, 1 m wide, and 1 m deep.*

You can easily find the volume of any large box or trunk by first
estimating or measuring its length, width, and depth, and then multiplying these dimensions. A handy approximate reference is a good-sized
refrigerator:

- A box fit around a good-sized refrigerator would have a
  volume of approximately 1 m^3.

- The inside of this good-sized refrigerator has a capacity
  of 0.5 m^3.

* In metric countries, cordwood is measured in a unit called the stere,
which is equal to 1 m^3.

4. What is the capacity of a closet 2 m wide, 2.5 m long, and 1 m deep? (How much can you store in this closet?) _____

_ _ _ _ _ _ _ _ _ _ _ _ _ _ _

$5 m^3$ (1 meter × 2 meters × 2.5 meters = 5 cubic meters)

5. See if you can supply all the following equivalents:

   (a) A good-sized refrigerator has a storage capacity of _____ cubic meter(s).

   (b) A tumblerful is _____ milliliter(s).

   (c) A tablespoonful is _15_ milliliter(s).

   (d) A pint basket holds _0.5_ liter(s) of berries.

   (e) A teaspoonful is _5_ milliliter(s).

   (f) A box 4 meters wide, 2 meters long, and 3 meters deep holds _____ m^3.

   (g) A quarter of a carton of milk holds _0.25_ liter(s).

   (h) A box into which a good-sized refrigerator fits exactly would hold _____ m^3.

_ _ _ _ _ _ _ _ _ _ _ _ _ _ _

(a) 0.5 cubic meter; (b) 250 milliliters; (c) 15 milliliters; (d) 0.5 liter; (e) 5 milliliters; (f) 24 m^3; (g) 0.25 liter; (h) 1 m^3

6. Suggest something with which you can associate each of the following volumes:

   (a) 1000 milliliters _1 liter = 1 qt = ? 4 cups_

   (b) 15 milliliters _1 tbl_

   (c) 1 m^3 _____

   (d) 500 milliliters _2 cups or 1 pt_

   (e) 1 liter _1 qt_

   (f) 0.25 liter _1 cup ?_

   (g) 4 liters _a gallon_

   (h) 0.5 liter _one pint_

   (i) 1 milliliter _____

(j) 2000 milliliters *half gallon*

(k) 250 milliliters *1 cup 8 ozs*

(l) 0.5 m³ _____

- - - - - - - - - - - - - -

(a) a quart carton; (b) 1 tablespoonful; (c) a box that will fit closely around a good-sized refrigerator; (d) a pint carton; (e) a quart carton; (f) a half-pint bottle (or a tumblerful or a glassful); (g) a gallon jug; (h) a pint carton; (i) the tip of the smallest finger, down to the beginning of the fingernail; (j) a half-gallon carton; (k) a half-pint bottle (or a tumblerful or a glassful); (l) the storage capacity of a good-sized refrigerator

METRIC COOKBOOK

Now you're ready to begin collecting recipes for your own metric cookbook. Here is a recipe to start you off:

Metric Salad

    1 bunch watercress
    250 liters mushrooms
    250 liters red kidney beans, drained
    1 Bermuda onion
    15 milliliters olive oil
    5 milliliters red wine vinegar
    1 milliliter garlic powder
    2 milliliters dry oregano
    1 milliliter powdered rosemary
    0.01 milliliter dry mint powder
    200 milliliters croutons

Chop vegetables in fine pieces; add beans. Combine olive oil, wine vinegar, and seasonings. Mix well. Toss salad together with dressing. Cool and serve with croutons on top.

CHAPTER FIVE
Weight and Mass

We love a measure
With metered pleasure,
And how we treasure
The milligram

But no one employs
The avoirdupois,
And nothing annoys
As much as the dram.

A familiar figure at amusement parks and country fairs is the fellow who guesses people's weight. He boasts that he can tell anyone's weight within 5 pounds, or forfeit a prize:

Step right up, folks, and let Bobo tell you what you weigh . . . . Your pick of a hundred beautiful prizes if he's off by more than 5 pounds.

Whether he guesses right or not, Bobo never loses. He charges at least twice what his giveaways cost. Yet the customer always goes away happy because he doesn't really look his weight—even to an old pro like Bobo.

Weights, like volumes, are deceptive. Nobody can tell exactly what something weighs just by looking at it. Bobo would be a real attraction —and show real skill as well as strength—if he could base his prize on a heft. Of course, he could still make a token guess first to flatter his victim.

See for yourself that weights are deceptive. Compare three look-alike vases—one made of iron, one of pottery, and one of plastic. The only way to tell what each one weighs is to actually lift it.

Here you'll learn to associate the weights of common objects with metric weights. The practice you'll get should give you a good sense of what the metric units weigh—probably a much better sense than you have now with units in the North American system. But before you're ready for that practice, you should have some background.

The two physical quantities <u>mass</u> and <u>force</u> are easily confused. The reason is that <u>weight</u> and <u>mass</u> are easily confused.

If you were an astronaut in space, the things around you would have very little weight. If you held a brick in the air and released it, it wouldn't fall; it would remain suspended in space. And if you placed the brick on a weight scale, it would probably register zero. That's because you would be so far away from the earth that you escape the pull of its gravity. It is the force of that gravity which, like a magnet,* pulls things downward and gives them weight.

<u>Weight</u>, then, is a measure of the force of a pull—gravitational pull. Even when you go up on a mountain, things weigh less because you are farther from the center of earth.

But the brick never loses its <u>mass</u> (or its matter). Although it wouldn't weigh anything in space, you could still stub your toe on it. And the more mass the brick has, the harder you would stub your toe. In other words, neither feather nor brick weighs anything in space. But you wouldn't hurt your toe on the feather because it has so little mass, and you would hurt it on the brick.

In science, the distinction between mass and weight is very important. But here on the surface of the earth and in everyday life, the weight of an object is proportional to its mass (i.e., very nearly the same as the mass). For most practical purposes, then, we can estimate the mass of an object by weighing it.

In Chapter 7, you will learn more about <u>force</u>. For now you may think of it as "the strength of push or pull." That strength is measured in SI by a unit called the <u>newton</u>. Weight, however, is a special kind of force, which we often use to estimate the size of mass in the SI unit of kilogram.

UNITS OF WEIGHT

Remember, weight is a force—actually a measure of the force of gravity. There are actually four major metric units of weight:

- the milligram (mg);
- the gram (g), which is the weight of 1 ml of water;
- the kilogram (kg), which is the weight of 1 $\underline{l}$ of water; and
- the metric ton** (t), which is the weight of 1 m^3 of water.

The milligram is so light (a grain of salt weighs about 1 milligram) it is seldom used except in medicine and other scientific areas. At the other extreme is the metric ton, closely equivalent to the ton we use now

---

* Gravity, however, is not a magnetic force.
** This is also spelled "tonne."

(2000 pounds). Very heavy things, like cars or industrial machinery, will be measured in metric tons. For some notion of how heavy a metric ton is, think of a box 1 meter wide, 1 meter long, and 1 meter deep. This box will hold 1 m^3. If you were to fill this box with water, the weight of this volume of water would be 1 metric ton, or 1000 kilograms.

The two units of weight we will use every day are the gram and the kilogram. Most dry packaged goods like rice and breakfast cereals will be measured in grams. Most canned goods will be measured in grams, too. In fact, if you check your pantry, you'll find that some manufacturers already list the weights of their products in grams as well as in pounds and ounces. Can you imagine the revolution in the supermarket when all canned goods of a similar size are in grams? With built-in unit-pricing it will be so easy to tell whether a bargain is really a bargain that all producers might even be encouraged to be honest!

Larger foodstuffs, like meats, will probably be measured in kilograms. A kilogram is equivalent to a thousand grams. (And 1000 kilograms is equal to 1 metric ton.) People's weights will also be measured in kilograms.

1. From this brief introduction, which unit do you think would probably be most useful for measuring each of the following items?

    (a) a button _____

    (b) a deck of cards _____

    (c) an elephant _____

    (d) a desk _____

– – – – – – – – – – – – – – –

(a) gram; (b) gram; (c) metric ton; (d) kilogram

THE GRAM

The best way to develop a good sense of weights is to actually lift different objects. This kind of experience is possible with things measured in grams or a few kilograms.

You'll have to do some preparation first. (If you were planning to work the exercises on weight while riding a bus or a train, please note that this one is not for the road.) For the gram, you should have these items at hand:

    5 straight pins
    5 nickels
    1 ordinary flashlight battery, size D (It is best to use one with
       a metal case.)

2 cans of tuna fish (7 oz)
1 telephone receiver

As you look over the following table of gram weights, lift each object or group of objects so you can get a good idea of what the weight feels like. (You can get a particularly good notion of the weight of objects measured in grams because they are so easy to heft.) Then try to associate each object or group with its weight in grams. The memory aids should help. You might also compare two similar weights by lifting one in your right hand and the other in your left.

Gram Weights

Object	Memory Aid	Weight
	5 straight (and "pointed") pins weigh "point 5 gram."	0.5 gram
	A nickel weighs 5 grams.	5 grams
	Five nickels weigh 25 grams.	25 grams
		100 grams
	A small can of tu-na weighs 250 grams.	250 grams

Object	Memory Aid	Weight
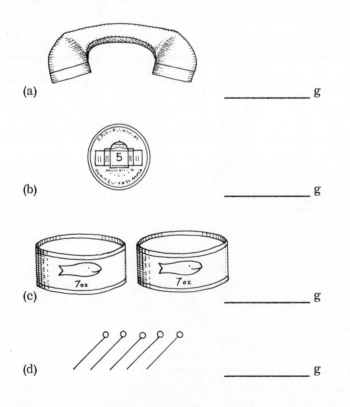	A telephone re-ceiver is in 3 parts.	300 grams
	<u>Two</u> cans of <u>tu</u>-na weigh 500 grams.	500 grams

2. Try to give the weight of each of these objects (or group of objects) without looking back at the table. You may want to lift the objects before you answer.

(a) _____ g

(b) _____ g

(c) _____ g

(d) _____ g

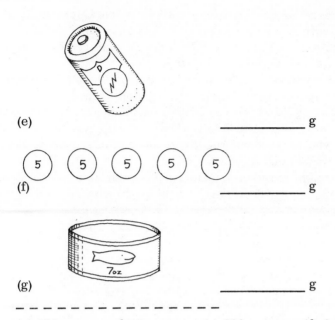

(e) ＿＿＿＿＿＿＿ g

(f) ＿＿＿＿＿＿＿ g

(g) ＿＿＿＿＿＿＿ g

- - - - - - - - - - - - - - - - - -

(a) 300 grams; (b) 5 grams; (c) 500 grams; (d) 0.5 gram; (e) 100 grams; (f) 25 grams; (g) 250 grams

3. Identify an object (or group of objects) that represents each of the following weights. Lift the items once or twice if you're not sure.

(a) 25 grams ＿＿＿＿＿＿＿＿＿＿＿＿＿＿＿

(b) 500 grams ＿＿＿＿＿＿＿＿＿＿＿＿＿＿＿

(c) 100 grams ＿＿＿＿＿＿＿＿＿＿＿＿＿＿＿

(d) 0.5 gram ＿＿＿＿＿＿＿＿＿＿＿＿＿＿＿

(e) 250 grams ＿＿＿＿＿＿＿＿＿＿＿＿＿＿＿

(f) 300 grams ＿＿＿＿＿＿＿＿＿＿＿＿＿＿＿

(g) 5 grams ＿＿＿＿＿＿＿＿＿＿＿＿＿＿＿

- - - - - - - - - - - - - - - - -

(a) 5 nickels; (b) 2 cans of tuna fish; (c) 1 battery, size D; (d) 5 straight pins; (e) 1 can of tuna fish; (f) a telephone receiver; (g) 1 nickel

THE KILOGRAM

And now for the heavyweights, measured in kilograms. For this exercise you will need a carton of milk (approximately 1 liter) and 1 six-pack

of canned beer or canned soda. A liter of milk is a convenient object to
associate with 1 kilogram, and a six-pack of beer or of soda weighs
about 2.5 kilograms (or 2500 grams). Lift the milk and then the six-pack
to get a better idea of how heavy each one feels.

What about the heavier objects? No one is going to heft a 90-kilogram
package just to get a good sense of what 90 kilograms feels like. This is
why human weights are good reference points. A handy one to remember
is that a big, 187-centimeter man weighs about 100 kilograms. (Some
well-known people who weigh about 100 kilograms are quarterback Joe
Namath and boxer Muhammad Ali.)

Find your own weight in the table below and memorize it. This will
be a reference point for you. You might do the same with the weights of
a few people you know.

The Conversion of Pounds to Kilograms

Pounds	Kilograms	Pounds	Kilograms
100	45	205	93
105	48	210	95
110	50	215	98
115	52	220	100
120	54	225	102
125	57	230	104
130	59	235	107
135	61	240	109
140	64	245	111
145	66	250	113
150	68		
		255	116
155	70	260	118
160	73	265	120
165	75	270	122
170	77	275	125
175	79	280	127
180	82	285	129
185	84	290	132
190	86	295	134
195	88	300	136
200	91		

4. The table itself should give you another reference. Notice that the pounds increase by 5.

   (a) What happens to the kilograms with each 5-pound increase?

   _____

   (b) Without looking at the table, what is your weight in kilograms?

   _____

   - - - - - - - - - - - - - - - -

   (a) There is at least a 2-kilogram increase for every 5-pound increase.

   (b) Refer to the chart on page 58 to check your answer.

5. Here is still another reference point to help you remember the kilogram scale. Think of a big man who weighs 100 kilograms as "100 percent" and a small woman as equal to half his weight, or 50 kilograms. Then you can relate other sizes to these reference points.

100 kilograms   50 kilograms

With these reference points in mind, estimate how much each of the following objects weighs. You can pick from the assortment of weights listed on the right. (You might want to cross out each weight you pick.) The list continues on the next page.

_____ (a) A heavy suitcase   9 kg

_____ (b) A Christmas turkey   1 kg

_____ (c) A pineapple   125 kg

_____	(d) A heavyweight wrestler	45 kg
_____	(e) A jockey	2.5 kg
_____	(f) The average man	205 kg
_____	(g) The average woman	60 kg
_____	(h) A cubic meter of water	25 kg
_____	(i) A six-pack of canned beer	80 kg
		1000 kg
		200 g

– – – – – – – – – – – – – –

(a) 25 kg; (b) 9 kg; (c) 1 kg; (d) 125 kg; (e) 45 kg; (f) 80 kg;
(g) 60 kg; (h) 1000 kg; (i) 2.5 kg

6.   Try to answer the following questions without looking back at previous hints or at the tables. All of your answers would be approximate, of course.

(a)   About how much does a big strapping man weigh?   _____

(b)   What does a liter of milk weigh?   _____

(c)   What do you think a 235-pound package would weigh?

_____

(d)   What do you think a 250-pound package would weigh?

_____

(e)   What does a six-pack of canned soda weigh in kilograms?

_____

(f)   What does a six-pack of canned soda weigh in grams?

_____

(g)   How much does 8 m^3 of water weigh in kilograms?

_____

(h)   How much does 8 m^3 of water weigh in metric tons?

_____

– – – – – – – – – – – – – – – –

(a) 100 kg; (b) 1 kg; (c) 107 kg; (d) 113 kg; (e) 2.5 kg; (f) 2500 g;
(g) 8000 kg; (h) 8 metric tons

7. For a review of weights and volumes, fill in the blanks:

(a) A milliliter of water weighs ___*1*___ g.

(b) A dm^3 weighs _____ g.

(c) A m^3 of water weighs _____ kg.

(d) A telephone receiver weighs _____ g.

(e) A teaspoonful of water is ___*5*___ ml.

(f) A quart of milk is about ___*1*___ l.

(g) A liter of milk weighs about ___*1*___ kg.

(h) A nickel weighs _____ g.

(i) The tip of your little finger has a volume of about _____ ml.

(j) A cm^3 is equal to _____ ml.

(k) A cm^3 of water weighs about _____ kg.

(l) A large man weighs about _____ kg.

(m) A metric ton of water would fill _____ m^3.

(n) A straight pin weighs _____ g.

(o) You weigh _____ kg.

(p) Remember your body references for this one. A cube the width of your hand holds _____ l.

– – – – – – – – – – – – – – – – –

(a) 1 g; (b) 1000 g; (c) 1000 kg; (d) 300 g; (e) 5 ml; (f) 1 l;
(g) 1 kg; (h) 5 g; (i) 1 ml; (j) 1 ml; (k) 1 g; (l) 100 g; (m) 1 m^3;
(n) 0.1 g; (o) your own answer; (p) 1 l

METRIC COOKBOOK

Here is a recipe for spaghetti to add to your collection:

Spaghetti Metrica

1 kilogram ground beef
15 milliliters olive oil
0.5 liter chopped onions
250 milliliters chopped mushrooms
250 milliliters chopped green (bell) pepper

1. 35 kilograms tomatoes, canned
150 grams tomato sauce
75 grams pine nuts (optional)
20 milliliters dried oregano
15 milliliters cayenne pepper
5 garlic cloves, crushed
1 Bay leaf
1 milliliter lemon juice
Salt to taste (approximately 15 milliliters)

Brown meat in oil. Mix all the other ingredients (including the meat) in a 4-liter pot. Cover and bring to a boil. Reduce heat and simmer for 1 hour.

For a complete metric meal, serve with a liter of good chianti, 500 grams of French bread, and metric salad (see page 51).

CHAPTER SIX

Temperature

Farewell, O fragile Fahrenheit,
Who kept us warm each winter night.
The next to go may well be us,
So treat us kindly, Celsius.

Of all the classes of units, temperature is the one that everyone under-
stands best, especially the natural or ambient temperature (the temperature of the air surrounding us). We are all forced to experience a 70-
degree day or a 90-degree day or a 10-degree day. When all else fails,
we can still talk about the weather and be content to do nothing about it.

We also have a good sense of body temperature, even though most of
us are not concerned with this every day. A body temperature of 98.6°
Fahrenheit is normal and indicates good health. A person with a temperature of 100° has a low-grade fever. If his temperature rises to 101 or
102, he's quite sick. If his temperature is much over 105, he's near
death. Small differences have large significance. No wonder we understand what body temperature means.

Those of us who cook also understand oven temperatures and what
they mean. A "slow" oven is 200 to 250°F; a moderate oven is 300 to
350°F; and a hot oven is 400 to 450°F. Broiling temperatures are 500°F
or more.

Conversions from the Fahrenheit to the metric scale require cumber-
some arithmetic. Fortunately, you won't have to make these conversions
for everyday uses of temperature. What you will need is the same sense
of temperature in metric units that you have now for Fahrenheit temperatures. This chapter will provide that understanding.

As you may remember from physics, heat is generated by the motion
of molecules; everything has some quantity of heat. Even ice in the refrigerator has moving molecules that produce heat. Of course, as we
take heat out of an object the molecules move less and less. If we took
the temperature of an object that had no heat (the molecules had stopped
moving), we would get no reading. This zero reading is called <u>absolute
zero</u> temperature.

We don't have any experience with absolute zero temperature since on the Fahrenheit scale it would register -459.67°F. That's just about as cold as anything can possibly get.* The zero point of the Fahrenheit scale, then, is just an arbitrary point. Gabriel Fahrenheit, a German who developed the mercury thermometer, experimented with temperature using a mixture of salt and ice. The coldest he could make the mixture he called "zero." (Of course, we know now that he was almost 460 degrees above the absolute zero.) As a result, on a Fahrenheit scale water freezes at +32°F and boils at +212°F. None of the convenient points (like 0, 10, 50, or 100) actually means anything.

A few years after Fahrenheit's scale was adopted, Ander Celsius, a Dane, suggested a scale on which zero would be the freezing point of water and 100 degrees the boiling point. These points are, of course, much more convenient than the Fahrenheit 32 and 212. Celsius's scale was adopted and has been the scale used in science and in the metric system. It used to be called the "centigrade" scale, because of its range from 0 to 100; now it is officially the Celsius scale—still °C— after its developer.

The diagram at the top of the facing page shows how the two scales compare.

More recently, science has adopted the Kelvin scale (°K). By starting at the "absolute" zero, the Kelvin scale has no minus (or negative) numbers. Its units are the same size as the Celsius scale. This means that a change of 1 degree on the Celsius scale is the same as a change of 1 degree on the Kelvin scale, as the second diagram on the facing page shows.

The degree-Kelvin is the official SI unit. Because the Celsius scale will be used in most practical situations, however, we will use the degree-Celsius as our unit—and it is acceptable in the SI system.

In the newly metric countries, people have simply had to start "thinking Celsius," since the arithmetic of converting from °F to °C is too complicated for daily use. Fortunately, it is easy to think Celsius:

        100°C and water boils.
        0°C and water freezes.

Memorize these points; you'll need them throughout the chapter.

---

* Technically, there is energy left even at -460°F, and the zero isn't as "absolute" as was once thought. But it is cold enough for us.

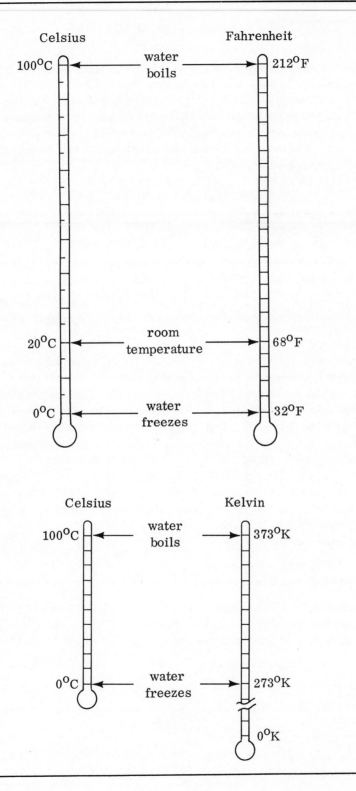

AMBIENT TEMPERATURE

First we're going to consider how to "think" ambient temperature on the Celsius scale. You've already learned that 100°C is boiling and 0°C is freezing. The table below will give you several reference points for temperatures in between. You'll note that there are memory aids for each reference point except for 20°C and -20°C. Study this table now. (You may, of course, substitute descriptions for the reference points that are more meaningful for you.)

Ambient Temperature

Temperature (°C)	Memory Aid	Reference Point
100		Boiling
40	Forty, Fiery	A hot, fiery day in summer in Arizona
30	Thirty, Thirsty	A thirsty, hot day in summer in New York
20		Room temperature
10	Ten, Tepid	A tepid day in autumn or in spring
0		Freezing
-20		A bitter cold day in winter
-40	Forty, Frigid	About as frigid as it ever gets in the United States

1. Suppose the temperature is above freezing but uncomfortably chilly. Suggest an approximate temperature that would fit this description.

_____

– – – – – – – – – – – – – –

about 5°C

2. How well do you recall the reference for ambient temperature?
Give the Celsius temperature that would approximately fit each of
these descriptions:

(a) A tepid day in May _____

(b) Freezing point _____

(c) A fiery day in August in Arizona _____

(d) As frigid as it gets in the U.S. _____

(e) Room temperature _____

(f) A very cold day in winter _____

(g) A thirsty, hot July day in New York _____

(h) Boiling point _____

_ _ _ _ _ _ _ _ _ _ _ _ _ _ _ _ _

(a) 10°C; (b) 0°C; (c) 40°C; (d) –40°C; (e) 20°C; (f) -20°C;
(g) 30°C; (h) 100°C

3. Suggest a fitting description for each of these temperatures:

(a) 0°C _____

(b) –40°C _____

(c) 20°C _____

(d) 100°C _____

(e) 30°C _____

(f) -20°C _____

(g) 40°C _____

(h) 10°C _____

_ _ _ _ _ _ _ _ _ _ _ _ _ _ _ _ _

(a) Freezing point; (b) As frigid as it ever gets in the U.S.; (c)
Room temperature; (d) Boiling point; (e) A thirsty day in New York
in the summer; (f) A very cold day in winter; (g) A fiery day in
summer in Arizona; (h) A tepid day in the spring or autumn

BODY TEMPERATURE

The table on the next page gives both Celsius and Fahrenheit temperatures and relates them to conditions of the body. Study these reference
points. It might help to first memorize the equivalent of a normal

temperature: $98.6^{O}F = 37^{O}C$. Notice that a person with the flu has a temperature of $39^{O}C$ and that a person convulses at a temperature of $41^{O}C$. Also keep in mind that a $1^{O}C$ rise is almost twice the temperature rise of $1^{O}F$, so far as its impact is concerned.

Body Temperature

Temperature (^{O}F)	Temperature (^{O}C)	Body Condition
98.6	37	Normal
99.5	37.5	Slight cold
100.5	38	Low-grade fever
102	39	Flu
104	40	Quite sick
105	40.5	Alarm
106	41	Convulsion

4. How would you describe the body condition of a person whose body temperature is $36.5^{O}C$?

   _____

   _ _ _ _ _ _ _ _ _ _ _ _ _ _ _

   below normal

5. What temperature in degrees Celsius would you expect for each of these body conditions?

   (a) An alarming condition _____

   (b) A slight cold _____

   (c) A convulsion _____

   (d) Flu _____

   (e) Normal _____

   (f) Very sick _____

   (g) A low-grade fever _____

   _ _ _ _ _ _ _ _ _ _ _ _ _ _ _

   (a) $40.5^{O}C$; (b) $37.5^{O}C$; (c) $41^{O}C$; (d) $39^{O}C$; (e) $37^{O}C$; (f) $40^{O}C$; (g) $38^{O}C$

6. A doctor takes his patient's temperature so he can get some idea of how sick the patient is. For practice, see if you can relate each temperature below to a body condition:

(a) 39°C _____

(b) 41°C _____

(c) 38°C _____

(d) 40.5°C _____

(e) 37°C _____

(f) 37.5°C _____

(g) 40°C _____

- - - - - - - - - - - - - - - - -

(a) Flu; (b) Convulsion; (c) A low-grade fever; (d) An alarming condition; (e) Normal; (f) A slight cold; (g) Quite sick

INDUSTRIAL AND KITCHEN TEMPERATURE

Let your own needs determine how you would treat this section. If you are a cook, you may want to memorize the kitchen temperatures. Otherwise, just read the exercise over to satisfy your curiosity about the temperatures in this higher range.

The boiling point for water is a good reference for industrial temperatures. At the other extreme is 3000°C, the temperature required to melt iron. The table below gives references for some industrial and kitchen temperatures. It continues on the next page.

Industrial and Kitchen Temperatures

Temperature (°C)	Heat Condition
3000	Melts iron
1000	Melts gold
250	For broiling steak
215	For roasting beef
200	For baking potatoes
175	For baking cookies
160	For baking fish

Temperature (^{O}C)	Heat Condition
150	For baking cake
125	For warming bread
100	For boiling water

7. You roast fowl at a temperature between that for baking potatoes and baking cookies—closer to the temperature for baking potatoes. Give an approximate temperature for roasting a turkey.

_____

– – – – – – – – – – – – – – – –

about 190°C

8. If you have memorized the references in the higher range of temperatures, try to give an approximate temperature for each of the following references. As a hint for some of them, keep in mind that 100°C —a basic reference point—separates the temperature for baking a cake and broiling a steak.

(a) Melts gold _____

(b) For baking potatoes _____

(c) Melts iron _____

(d) Boils water _____

(e) For broiling steak _____

(f) For warming bread _____

(g) For baking a cake _____

– – – – – – – – – – – – – – – –

(a) 1000°C; (b) 200°C; (c) 3000°C; (d) 100°C; (e) 250°C;
(f) 125°C; (g) 150°C

9. Skip this exercise if you haven't memorized the references for temperatures in the higher range. Otherwise, try to suggest a reference for each temperature given below and on the following page:

(a) 150°C _____

(b) 3000°C _____

(c) 250°C _____

(d) 125°C _____

(e) 200°C _____

(f) 1000°C _____

(g) 100°C _____

– – – – – – – – – – – – – – –

(a) For baking cake; (b) Melts iron; (c) For broiling steak;
(d) For warming bread; (e) For baking potatoes; (f) Melts gold;
(g) Boils water

CHAPTER SEVEN

Work, Power, and Other Quantities

Ohm, sweet Ohm

What of the watt as the horsepower's source,
With kilos and millis, the unit of power!
Behold the bold newton as a measure of force,
And the duel of the joule and the kilowatt-hour!

Hail pascal, that rascal, that prince of a measure!
Frown on the pound and the passing pound-force.
Let a jolt from the volt spoil faraday's pleasure,
And pamper the amp that lights up the course!

This chapter was designed for students and employees of science and technology, and also for those who find these topics interesting. The physical quantities discussed here do have some everyday use. We do pay for electric energy by the <u>kilowatt-hour</u> and buy light bulbs by the <u>watt</u>. The use of these two metric units shows how far metrication has gone in North America.

On the other hand, you'll have little daily need to think in metric units of force and pressure, energy and power. So, you don't need to memorize these units; just familiarize yourself with them.

Learning units of length, weight, volume, and temperature is like learning to speak and hear a new language fluently. But learning units of force and pressure (also force and energy) is like preparing yourself to read a new language with the use of a dictionary. The exercises in this chapter will give you this kind of practice.

FORCE

When you want to move an object, the force you would have to exert depends on two things:

- How much mass the object has. (You have to push harder to move a 1000-kg piano than a 500-kg piano.)

- How much you accelerate the object. (More force is required for a car to accelerate from zero to 20 km/h in a second than from zero to 10 km/h in the same period of time.)

To measure how much force one object exerts on another, both mass and acceleration must be considered.

The basic unit is defined as the amount of force required to accelerate 1 kilogram 1 meter per second per second. This unit of force is called a <u>newton</u> (N), after Isaac Newton, the English philosopher who created the calculus and who is often called the father of classical physics.

It's rather hard to get a feel for "accelerating a kilogram 1 meter per second per second (1 m/s/s)." Let's try to visualize this another way. If you hold a 100-gram object in your hand, the earth's gravity (at sea level) exerts a force of about 1 newton on this object. Or, to put it another way, you have to apply a force of 1 newton to hold a 100-gram object in the air.

Remember from Chapter 5 that the reference for 100 grams is a flashlight battery. If you heft this battery, you'll get the sense of a newton of force. An apple may also help you remember what a newton is. A small apple (which weighs about 100 grams) exerts a downward force of about 1 newton. The apple is a good memory aid, if you recall the story of Sir Isaac Newton's sudden discovery of the law of gravity when an apple fell on his head as he sat under a tree.

1. Complete the following sentences:

(a) The weight of a flashlight battery can be expressed as

_____ newton or _____ grams.

(b) A 100-kilogram man exerts a downward force of _____ N.

(c) A 25-kg child, riding piggyback, would exert _____ newtons of force on your back.

– – – – – – – – – – – – – – – –

(a) 1 N or 100 g; (b) 1000 N; (c) 250 N

2. Weight is one kind of force—the product of the mass of an object and the acceleration imparted by gravity. Technically, weight should be measured in newtons, since the pull of gravity varies from place to place. For example, weight is less on top of a mountain than at sea level. In much technical work in physics, weight is actually measured in newtons. However, for many purposes, and certainly for everyday use, weight is a useful way of estimating the mass of an object. At any particular place on earth, the acceleration of gravity will be constant. If two stones exert different downward forces, it must be because they have different masses. So, we can estimate the mass of an object if we measure the downward force of this object on a scale at a given place on earth.

Physicists have long used the dyne as a unit of very small force: 100,000 dynes is equal to 1 newton. However, the international agreement that created the SI units didn't include the dyne. In SI, units of small force are expressed as micronewtons (μN) and millinewtons (mN).

From what we've said about force, see if you can answer these questions:

(a) If you hold a 1-kg ball in your hand, approximately how many newtons (N) of force does it exert on you? _____

(b) An object exerting a force of 10 N downward on a scale would weigh _____ kg.

(c) A 1000-kg car exerts a force of _____ N on the earth.

(d) If we measured weights in newtons instead of in grams, a 100-newton man would weigh about _____ kilograms at sea level.

– – – – – – – – – – – – – – – –

(a) 10 N; (b) 1 kg; (c) 10,000 N; (d) 10 kg

PRESSURE

Pressure is a quantity related to force, but probably more familiar. In the filling station, for example, we ask for 28 pounds of air in the rear tires. What we are really asking for is an amount of air that is exerting 28 more pounds of force per square inch on the inside of the tire than the pressure of air on the outside of the tire.

Pounds per square inch (psi), the most common NAS unit of pressure, will be replaced by the newton per square meter or N/m^2. It will be called the <u>pascal</u>, after Blaise Pascal, the seventeenth-century French philosopher who contributed greatly to mathematics and engineering. The pascal, or Pa, is equal to a force of 1 newton exerted on an area of 1 square meter.

One pascal is not really very much pressure—not nearly as much as one psi. Since a pascal is the force of a newton spread over a square meter, you can get some idea of it if you spread half a cup (about 100 grams) of sugar evenly over the top of a table that is 1 meter on each side. The pressure of the sugar exerted at any one point on the table would be quite small. It would, in fact, take 200,000 pascals (N/m^2) to inflate an ordinary automobile tire. (Compare this with the 28 pounds.)

For larger pressures, we use a larger unit—meganewtons per square meter, or MN/m^2. (A meganewton, remember, is 1 million newtons.) Another even larger size is the giganewton per square meter, or GN/m^2. (A giganewton is 1 billion newtons.) There are also MPa (megapascals) and GPa (gigapascals).

Another common metric unit of pressure is the <u>bar</u> (b), which is equal to 100,000 Pa. The bar is particularly useful in meteorology.

Before continuing to the next exercise, you might want to reread these definitions.

3. Try these simple calculations of pressure:

   (a) How many bars of pressure would be needed to inflate an ordi-
        nary automobile tire? _____

   (b) If a MN/m^2 = 1,000,000 N/m^2, how many MPa will inflate an
        ordinary tire? _____

   _ _ _ _ _ _ _ _ _ _ _ _ _ _ _ _ _

   (a) 2 b; (b) 0.2 MPa

4. Complete the following statements about force and pressure by check-
    ing the correct answer:

   (a) The weight produced by gravity acting on an object is

        _____ (1) force
        _____ (2) pressure

   (b) Newtons per square centimeter would be a measure of

        _____ (1) force
        _____ (2) pressure

(c) A pascal is equal to a

_____ (1) newton
_____ (2) N/m^2
_____ (3) bar

(d) If a bar equals 100,000 pascals, 3 bars equal

_____ (1) 3 MN/m^2
_____ (2) 0.3 MN/m^2

(e) The pressure exerted by a 100-gram object is

_____ (1) 1 newton
_____ (2) 1 newton/m^2
_____ (3) depends on the area it covers

(f) A 100-gram object resting on a weight scale exerts a force of about

_____ (1) 1 newton
_____ (2) 1 pascal
_____ (3) 1 bar

(g) The pressure required to inflate a toy balloon would be about

_____ (1) 10,000 N   _____ (3) 0.1 bar
_____ (2) 10,000 Pa   _____ (4) 0.1 N

_ _ _ _ _ _ _ _ _ _ _ _ _ _ _

(a) 1; (b) 2; (c) 2; (d) 2; (e) 3; (f) 2; (g) 2 and 3

5. Fill in the blanks:

(a) Meteorologists would measure how "heavy" the air is in units
of _____.

(b) Although weight is often measured in grams, it can also be
measured in _____.

_ _ _ _ _ _ _ _ _ _ _ _ _ _ _

(a) bars; (b) newtons

WORK AND ENERGY

In physics, energy is defined as the capacity to do work. When you pay
your electric bill, you pay for kilowatt-hours—the energy the electric
company supplies to do work for you. When you buy a window air condi-

tioner, you buy Btu's—the amount of work this air conditioner will do to cool the air for an hour. When you count the calories you eat, you count units of food energy, some of which get stored in your body as fat for future work. Kilowatt-hours, Btu's, and calories are among the many units used to describe work and energy.

One reason for so many units is that energy comes in so many forms. One example is <u>heat energy</u>, the capacity of matter to do work as it burns. A Btu (British thermal unit) is defined as the amount of heat needed to raise 1 pound of water 1 degree Fahrenheit. Obviously, it isn't a metric unit. A calorie—or the amount of heat needed to raise a gram of water 1 degree Celsius—is a metric unit, although not the one adopted for use by SI. To get some idea of what a calorie is, consider that a teaspoonful of sugar has about 15 calories. If you burned it very efficiently, you would raise the temperature of 15 grams of water 1 degree Celsius.

<u>Mechanical energy</u>, another kind of energy, is measured in the NAS as foot-pounds (ft-lb). If you lift a 1-pound book 1 foot, you would perform 1 foot-pound of work.

<u>Electrical energy</u>, a third kind of energy, is measured in both the NAS and the metric system as kilowatt-hours (kwh). A kilowatt-hour is about how much energy you would use to burn a couple of light bulbs all night long. Electrical engineers, electricians, and electrical contractors won't have to learn a new measurement system because electrical units are universally metric.

One property of energy is that one kind of energy can be converted to another kind. This means that we can burn gas (heat energy) to turn an engine (mechanical energy). Or we can generate electricity by turning a wheel. The ordinary light bulb shows us that electrical energy gets converted to light energy and heat energy. Since energy is basically the same whether it is heat or mechanical or nuclear or electrical or light, it makes sense to have one system of units with as few different units as possible. Science and engineering have been greatly inconvenienced by the use of so many systems of units to measure work and energy.

Work can be defined as force exerted through a distance. An object resting on the palm of your hand exerts a force, but no work is done because nothing is moved.* But if you drop the object so that it falls on a spring, it will move the spring and thus do work. The farther the object falls (i.e., the greater the distance it moves through), the farther it will move the spring. <u>Work</u>, then, can be defined as <u>force times distance</u>.

---

* When you hold a heavy object, you may be doing "work" psychologically, but you are doing no physical work because you aren't moving the object.

If you exert a force of 1 newton through a distance of 1 meter, you will have done 1 metric unit of work. This unit is called a joule (J), which is pronounced JOOL, almost like "jewel." But that is a technical definition. A good way to get an idea of a joule of work is to lift something (work against the force of gravity). If you lift a flashlight battery (100 grams) 1 meter, this activity requires the energy of about 1 joule. If you then drop the flashlight battery, its impact on the floor will yield 1 joule of energy. If you drop the battery 1 meter and it strikes a spring, it would do 1 joule of work in moving the spring. Of course, if you moved the battery horizontally, it would take less work since you would not be opposing gravity.

You might want to review these definitions before continuing.

6. See if you can answer these questions without looking back:

(a) If you exert a force of 10 N through a distance of 10 meters, how much work would you do? _____

(b) If you lift 1 kg a distance of 1 meter, how much work would you do? _____

(c) If you lift 10 kg a distance of 10 meters, how much work would you do? _____

(d) If you lift a heavy man 1 meter, about how much energy would you expend? _____

(e) Do you remember the mass of a nickel in grams? How much energy does it require to lift 20 nickels 1 meter?

_____

– – – – – – – – – – – – – – – –

(a) 100 J; (b) 10 J; (c) 1000 J; (d) 1000 J; (e) 1 J

7. Another metric unit of energy is the erg, which is equal to one ten-millionth of a joule, or about the energy required to lift a straight pin 1 centimeter. Since the erg is not accepted as an SI unit, small amounts of energy can be expressed as millijoules (mJ).

Presumably, the joule will replace both the calorie and the erg as a measure of energy. A calorie is just a little more than 4 joules. At least you can think you are eating more when your doctor restricts your diet to 8000 joules a day (rather than 2000 calories). A teaspoonful of sugar contains about 100 joules of energy. If we burned that much sugar, we would release about that much energy.

Try answering these questions about energy:

(a) The energy released when we burn a teaspoonful of sugar is approximately equal to the energy required to lift _____ flashlight battery (or batteries) a distance of 1 meter.

(b) A joule is equal to the force of 1 _____ applied through a distance of 1 _____ .

(c) Lifting a heavyweight boxer 0.5 meter would require how much energy? _____ J

- - - - - - - - - - - - - - - -

(a) 100; (b) newton, meter; (c) 500 J

POWER AND ENERGY

If a small child carries a load of bricks one brick at a time from one place to another, he does just as much work as a strong man who carries the same load in one trip. Both are moving the same number of kilograms through the same distance. Although the strong man does the same work, he works at a faster rate. This rate or speed of work is called power.

Just as velocity is measured as meters per second or kilometers per hour, power is measured as work per second, or joules per second. If it takes me twice as long to do a joule of work as it takes you, you have twice my power even though we may use the same energy.

8. In each of the problems below, determine who does more work and who exerts more power—Mr. X or Mr. Y.

(a) Mr. X carries 100 kg a distance of 1 km in an hour. Mr. Y carries 200 kg for 1 km in 10 hours. Who does more work?

_____ Who exerts more power? _____

(b) Mr. X carries 1 kg for 1 meter in 1 second. Mr. Y carries 2 kg for 1 meter in 5 seconds. Who does more work?

_____ Who exerts more power? _____

- - - - - - - - - - - - - - - -

(a) Y, X; (b) Y, X

9. In sports, the difference between power and energy is dramatic. There is the skinny fellow who runs 3000 meters in 500 seconds (8 minutes), and the strapping fellow with great leg muscles, who

runs 100 meters in 10 seconds. Check which one does more work, and which one exerts more power.

(a) More work? _____ (1) 3000-meter man
_____ (2) 100-meter man

(b) More power? _____ (1) 3000-meter man
_____ (2) 100-meter man

_ _ _ _ _ _ _ _ _ _ _ _ _ _ _ _

(a) 1; (b) 2

10. The long-distance runner obviously expends more energy than other runners since he's carrying roughly the same mass through a greater distance. (The amount of work doesn't depend on the time it takes.) But the 100-meter-dash man moves his weight 10 meters per second (100 meters divided by 10 seconds), whereas the distance runner moves at 6 meters a second (3000 meters divided by 500 seconds). So the sprinter exerts more power than the distance runner.

(a) Suppose you had to choose a participant in a sporting event requiring moderate power but great energy (like tennis). Who would you choose?

_____ (1) a 60-kg woman athlete in peak condition
_____ (2) a 120-kg man in poor condition and gone to flab

(b) Suppose you had to choose a participant in a sporting event requiring great power but moderate energy (like weight-lifting). Who would you pick?

_____ (1) the woman athlete
_____ (2) the poorly conditioned man

_ _ _ _ _ _ _ _ _ _ _ _ _ _ _ _

(a) 1; (b) 2

11. In sports "strength" means power or the ability to do work rapidly. "Stamina," on the other hand, means the ability to expend great energy over longer periods of time.

An electric power company also carefully distinguishes between power and energy. And it must build an electric power plant for two different purposes: to supply power and to supply energy.

Suppose a resident in each of the 100 houses in a town used a 1000-watt clothes iron for 100 seconds a day, but everyone used his iron at a different time. The electric company would have to provide 1000 watts times 100 seconds times 100 houses, or 10 million units (joules) or energy. Its generators would have to run long enough to supply it. But the power company would have to generate only 1000 watts of power since that is the largest power demand at any given second.

Now suppose all the townspeople began to use their irons at the same time. Then the power company would have to provide the same amount of energy (10 million joules). But since it must produce all that energy during a short time (1 second), it would have to generate 1000 watts times 100 houses, or 10,000 watts of power. To do this, the power company would have to buy larger (more powerful) generators.

Most electric power companies have two charges for their services—an energy charge and a demand (power) charge. From what you have read here, but without looking back, see if you can complete these statements.

(a) If a factory does relatively little electrical work but uses its electricity all at once, it will have

_____ (1) a big power (demand) charge
_____ (2) a big energy charge

(b) A factory that uses moderate amounts of electricity steadily around the clock will have

_____ (1) a big power charge
_____ (2) a big energy charge

– – – – – – – – – – – – – – – – –

(a) 1; (b) 2

12. The SI unit for power is the watt (W), named after James Watt, the eighteenth-century Scotsman who invented the modern steam engine. The watt is equal to 1 joule per second, or 1 W = 1 J/s. Lifting a flashlight battery 1 meter equals a joule of energy; you will use 1 watt of power if you lift the battery that far in 1 second.

(a) If you expend 1 joule of energy in 4 seconds, how much power

do you exert? _____

(b) If you perform 10 joules of work in 2 seconds, how much power

is required? _____

– – – – – – – – – – – – – – – –

(a) 0.25 W; (b) 5 W

13. It doesn't take a tremendous amount of power to do a joule of work in 1 second. For larger units of power, the kilowatt (kW) is commonly used.

(a) A 100-watt light bulb requires _____ joules per second.

(b) A dishwasher motor that performs 1000 joules per second would

   be rated as requiring _____ W or _____ kW of power.

(c) A 1000-kilowatt engine is capable of doing _____ joules per second of work.

– – – – – – – – – – – – – – – – –

(a) 100 J/s; (b) 1000 W or 1 kW; (c) 1,000,000 J/s

14. The metric unit of power that is used depends upon the amount of power being measured. Watts are used for measuring intermediate-sized power, like that required to operate home appliances. Milli-watts and microwatts are used to measure the tiny power require-ments in electronic equipment. When conversion to the metric system is complete, the kilowatt will replace many larger units of power, including the British thermal unit (Btu), horsepower, Btu per second, and the kilocalorie per minute. The kilowatt will be used, among other things, to measure the mechanical and electrical power output of engines and generators and the heat demand of build-ings.

Electric power companies use the kilowatt as a measure of the power they supply their customers. But the energy these companies supply is so great that billions and billions of joules would be required to measure it. Even the kilojoule is too small a measure for the energy the power company produces. To have a large-enough energy unit and to keep the same prefix used to measure power (kilo), the electrical industry uses a unit called the kilowatt-hour. Your elec-tric bill, in fact, is measured in this unit. One kilowatt-hour (kW-h) equals 3,600,000 joules. Here is how to figure the kilowatt-hour:

   1 kilowatt = 1000 joules per second

   Multiply both sides of this equation by 1 hour (3600 seconds):

   kW × h = 3600 seconds × 1000 joules
   1 kW-h = 3,600,000 joules

There is a way to visualize a kilowatt-hour. If lifting a flashlight battery 1 meter expends 1 joule of energy, then to perform 1 kW-h of work, you would have to lift 3.6 million batteries 1 meter (or one battery 3600 kilometers). Or, in terms of electric light, one kW-h of electrical energy will operate twenty 50-watt light bulbs for an hour.

Or, the engine of a small American car can generate about 100 kilowatts of power. A tiny foreign car engine generates about 50 kilowatts of power and a large American car engine generates about 250 kilowatts. Try to replace horsepower with kilowatt in your thinking. One kilowatt equals about three-quarters of a horsepower.

You probably won't be able to remember all of these units or their relationships to other units. But you should have some idea now of what they are.

(a) If the engine of a small American car went full throttle for one

hour, how much energy would it use? _____

(b) If a 100-watt bulb burned for 0.01 second, how much energy

would it consume? _____

‒ ‒ ‒ ‒ ‒ ‒ ‒ ‒ ‒ ‒ ‒ ‒ ‒ ‒ ‒ ‒

(a) 100 kW-h; (b) 1 J

15. The table on page 85 should help you master the distinction between units of force, pressure, energy, and power. Study the table and then answer the questions below.

(a) If you want to know whether or not an engine is large enough to pull a 1000-kg wagon, you would ask about its rating in

_____ (1) pascals
_____ (2) kilowatt-hours
_____ (3) kilowatts

(b) If you pumped water for people, you would charge them by

_____ (1) newtons
_____ (2) kilojoules
_____ (3) kilowatts

(c) If you measured the difference between the weight of an astronaut on the earth and on the moon, you would be measuring in

_____ (1) newtons
_____ (2) pascals
_____ (3) kilograms

(d) Suppose you decide to charter a plane costing $200 per hour rather than a smaller one costing $100 per hour. What would you be getting for the extra $100?

_____ (1) more newtons
_____ (2) more joules
_____ (3) more watts

– – – – – – – – – – – – – – – –

(a) 3; (b) 2; (c) 1; (d) 3

16. Study the table on page 85 once more. Then try to work the following problems without looking at the table.

(a) How much work would be required to lift a nickel 20 meters?

_____

(b) How much power would be required to operate an iron that consumes 1000 joules of energy per second? _____

(c) How much energy would the same iron consume if it were used for 1 hour? (Give your answer in a unit other than the joule.)

_____

(d) A 1-kilogram box resting on the table exerts a downward force of about _____.

(e) A square centimeter is 1/10,000 the area of a square meter. How many N/cm^2 would inflate a tire? _____

(f) The engine of a Cadillac automobile is likely to have a rating of

_____ (1) 50 kW
_____ (2) 250 kW
_____ (3) 50 kW-h
_____ (4) 250 kW-h

– – – – – – – – – – – – – – –

(a) 1 J; (b) 1 kW; (c) 1 kW-h; (d) 10 N; (e) 20 N/cm^2; (f) 4

	Force	Pressure	Energy	Power
Basic Unit Symbol	N (newton)	Pa (pascal)	J (joule)	W (watt)
Other Common SI Units		b, N/m^2 (bar)	kW–h (kilowatt–hour)	kW (kilowatt)
Equivalence		1 b = 100,000 Pa		1 kW–h = 3.6 million J
Relationship to Energy	N = J/m	Pa = J/m^3		W = J/s
Reference Sizes	Weight of a flashlight battery = 1 N	Pressure to inflate an auto tire = about 0.2 MPa Pressure to inflate a toy balloon = 10 kPa	Lifting a flashlight battery 1 meter = 1 J Burning ten 100-watt bulbs for an hour = 1 kW–h	Lifting a flashlight battery 1 meter in 1 second = 1 W Power of a small American car engine = 100 kW

CHAPTER EIGHT

Converting to Metric Units

A thorn by any other length is still a thorn;
No rule-of-thumb will heal the flesh that it has torn.

If you're really thinking metric, you'll have no reason to convert from
metric to NAS units. In fact, you should avoid converting from the
metric system in any case, because it will hinder your thinking metric.
We feel so strongly about it that we do not discuss converting from
metric to NAS units.

However, since it is so much easier to think metric than it is to
think inch or pound, there is good reason to convert pounds to kilograms
and inches to centimeters. Any carpenter, for example, would be foolish
not to take advantage of metric simplicity.

Practice you get in converting from NAS to metric units won't inter-
fere with your thinking metric. On the contrary, it can be a great learning aid. Unfortunately, you'll soon forget what you've learned in this
book unless you practice. And this is true no matter how well the writers
have done their jobs or how good a student you have been. But if you
learn and practice using the metric "rules-of-thumb" for conversions in
section A of this chapter, you will become a seasoned pro.

The exact conversion factors are hard to remember. An inch, for
example, equals 2.54 centimeters; a pound is 0.4732 liter. But if you
do need to make a conversion for some technical use, precision will
usually be important. You can then use the tables in the Appendix. For
most purposes, however, our rules-of-thumb will be accurate enough.

A: RULE-OF-THUMB CONVERSIONS

The rules-of-thumb for conversions are reasonably easy to memorize.
Even so, it will help greatly if you will understand and remember this
rule:

> When converting from a larger to a smaller unit, you get
> a larger number.

For example, if you want to convert 10 inches to centimeters, multiply
the 10 by 2.5:

Larger Unit		Smaller Unit
10 inches	=	25 centimeters

This rule will be obvious when you think about it, but it is a little tricky
to remember. The diagram below illustrates further the rule that when
converting from a larger unit (here, inches) to a smaller unit (centimeters), you get a larger number:

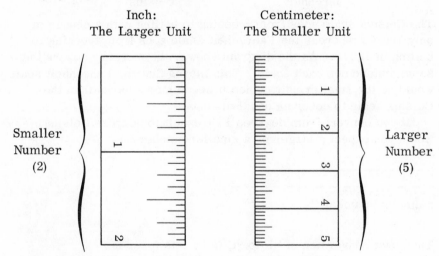

Inch:
The Larger Unit

Centimeter:
The Smaller Unit

Smaller
Number
(2)

Larger
Number
(5)

Obviously, the reverse of this rule is also true:

> When converting from a smaller to a larger unit, you get
> a smaller number.

Temperature Conversions

1. The first place we can use our rule to advantage is in converting
    from degrees Fahrenheit to degrees Celsius. Let's compare the
    two temperature scales in the diagram on the next page.

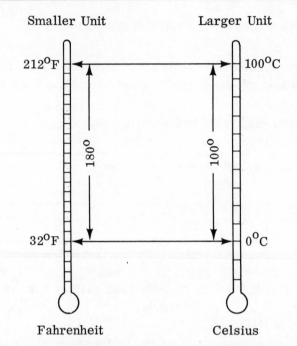

Smaller Unit   Larger Unit

212°F 180°   100°C 100°

32°F   0°C

Fahrenheit   Celsius

The Celsius scale goes from freezing to boiling with a change of
only 100°C, whereas the Fahrenheit scale goes from freezing to
boiling in 180°F. As the diagram shows, the mercury covers the
same distance on each scale. This means that the Fahrenheit scale
would be the smaller unit, since it needs more marks than the
Celsius scale to cover an equal distance.

If you convert from degrees Fahrenheit to degrees Celsius,
would you expect a larger or a smaller number?

_ _ _ _ _ _ _ _ _ _ _ _ _ _ _

a smaller number

2.   The exact conversion of °F to °C is by this formula:

$$°C = \frac{5}{9} (°F - 32)$$

The conversion of a temperature of 122°F goes like this:

$$°C = \frac{5}{9} (122°F - 32)$$

$$= \frac{5}{9} (90)$$

$$= 50°C$$

You'll agree this is a clumsy conversion.

Fortunately, you can make a reasonably accurate conversion in your head and without the formula. The rule is simply this:

$$^\circ C = \tfrac{1}{2}\,^\circ F - 15$$

In other words, if you take half the temperature in $^\circ F$ and subtract 15, you'll have a close estimate of $^\circ C$—at least, close enough for everyday use.

The table below compares some exact conversions with the rough ones and points out the amount of error:

$^\circ F$	Exact Conversion $^\circ C$	Rough Conversion $^\circ C$	Error $^\circ C$
100	37.78	35	2.78
50	10	10	0
32	0	1	1
0	−17.77	−15	2.77

If we made the conversion of $32^\circ F$ to $^\circ C$ using the exact temperature conversion formula, we would get 0. What amount of error would we get if we used the rough conversion instead?

_____

— — — — — — — — — — — — — —

$1^\circ C$

3. Now you can practice this simple conversion in your head. Remember the rule: Halve the Fahrenheit and subtract 15.

(a) $110^\circ F =$ _____ $^\circ C$

(b) $60^\circ F =$ _____ $^\circ C$

(c) $-10^\circ F =$ _____ $^\circ C$

(d) $84^\circ F =$ _____ $^\circ C$

(e) $90^\circ F =$ _____ $^\circ C$

(f) $8^\circ F =$ _____ $^\circ C$

— — — — — — — — — — — — — —

(a) $40^\circ C$; (b) $15^\circ C$; (c) $-20^\circ C$ (Watch your signs when you subtract!); (d) $27^\circ C$; (e) $30^\circ C$; (f) $-11^\circ C$

4. Before we continue, make certain that you don't get into the habit of converting rather than thinking metric directly. To review the Celsius scale, then, complete the following sentences:

   (a) Room temperature is _____ °C.

   (b) A hot July day in New York City is _____ °C.

   (c) A scorching day in the desert is _____ °C.

   (d) Freezing is _____ °C.

   (e) A bitter cold winter day is _____ °C.

   (f) The coldest day of the year in North Dakota is _____ °C.

   (a) 20°C; (b) 30°C; (c) 40°C; (d) 0°C; (e) –30°C; (f) –40°C

5. Make the simple conversion for these temperatures without looking back at the previous exercise:

   (a) 50°F = __*10*__ °C

   (b) 40°F = __*5*__ °C

   (c) 70°F = __*20*__ °C

   (d) 30°F = __*0*__ °C

   (e) 20°F = __*–5*__ °C

   (f) 100°F = __*35*__ °C

   (a) 10°C; (b) 5°C; (c) 20°C; (d) 0°C; (e) –5°C; (f) 35°C

6. This simple way to convert temperatures works well for ambient temperatures and for temperatures up to the boiling point of water. For temperatures at and above the boiling point of water, there is an even simpler rule: Halve the °F value. The table below shows how the exact conversion compares with the rough one:

°F	Exact Conversion °C	Rough Conversion °C
212	100	106
482	250	241
932	500	466

The rough estimate is not too far off, as you see.

Choose the appropriate rule and then make rough conversions for the following temperatures:

(a) 600°F = _____ °C

(b) 260°F = _____ °C

(c) 120°F = _____ °C

(d) 800°F = _____ °C

(e) 200°F = _____ °C

(f) 160°F = _____ °C

– – – – – – – – – – – – – – – –

(a) 300°C; (b) 130°C; (c) 45°C; (d) 400°C; (e) 85°C; (f) 65°C

Length: Inches to Centimeters

7.   Before you practice converting from inches to centimeters, you would do well to refresh your metric thinking. Begin by recalling your references for length. What are your basic references for these lengths?

(a) 1 mm _____

(b) 10 mm (1 cm) _____

(c) 25 mm (2.5 cm) _____

(d) 50 mm (5 cm) _____

(e) 100 mm (10 cm) _____

(f) 200 mm (20 cm) _____

(g) 500 mm (50 cm) _____

(h) 1000 mm (100 cm) _____

(i) 1500 mm (150 cm) _____

Draw a line for each of the following lengths:

(j)   12.5 mm

(k) 2 cm

(l)   6 cm

(m) 8 cm

– – – – – – – – – – – – – – – –

(a) - (i) Check your answers by turning to the table on pages 21-22.
(j) - (m) Check your answers by measuring your lines with a metric ruler.

8. The multiple 2.5 is an accurate-enough estimate to use for converting inches to centimeters.

$$10 \text{ inches} = 2.5 \times 10 \text{ inches} = 25 \text{ centimeters}$$

You'll find it easier to remember this factor if you develop the sense of an inch in centimeters. A section of one of your fingers is just about an inch long—probably the mid-section of your little finger.

2.5 cm

Measure and find a reference that works for you.

If you compare your centimeter reference with your inch reference, you'll see that a span of 2.5 centimeters makes an inch. Of course, 2.5 is troublesome as a multiplier—if you do the multiplying on paper. It's fairly easy if you do it in your head. You can use this rule:

Double and add half.

To convert 40 inches to centimeters, you would multiply 2.5 by 40. Or, using the double-and-add-half rule:

double 40 = 80
add half, or 80 + 20 = 100

Don't look back. Use this rule and convert (roughly) the following inches to metric units. Watch your prefixes!

(a) 60 in. = _____ cm

(b) 220 in. = _____ mm

(c) 30 in. = _____ cm

(d)   420 in. = _____ mm

(e)   640 in. = _____ m

(f)   120 in. = _____ cm

– – – – – – – – – – – – – – – – –

(a) 150 cm; (b) 550 mm; (c) 75 cm; (d) 10,500 mm; (e) 16 m;
(f) 300 cm

Length: Miles to Kilometers

9.   Complete the following sentences to review your metric sense.
(Remember, think of thruway speed as 100 km/h.)

(a) Speed on a back road is _____ km/h.

(b) City driving is _____ km/h.

(c) School–zone driving is _____ km/h.

(d) Jogging is _____ km/h.

(e) Walking is _____ km/h.

– – – – – – – – – – – – – – – – –

(a) 75 km/h; (b) 50 km/h; (c) 25 km/h; (d) 10 km/h; (e) 5 km/h

10. A mile is equal to 1.609 km. This suggests that a simple rule for
conversion would be:

     Add 60 percent.

In other words, 100 mph is the same as 100 mph + 60% of 100 mph,
or 160 km/h.
     Now do these conversions, using this rule: Add 60 percent.

(a)   200 miles = _____ km

(b)   50 miles = _____ km

(c)   6 miles = _____ km

(d)   400 miles = _____ km

(e)   30 miles = _____ km

(f)   12 miles = _____ km

– – – – – – – – – – – – – – – – –

(a) 320 km; (b) 80 km; (c) 9.6 km; (d) 640 km; (e) 48 km;
(f) 19.2 km

11. Let's review all the rough conversions for length and temperature.
    Remember the rules, and watch your prefixes!

(a) 6 inches = _____ cm

(b) $70^{\circ}F$ = _____ $^{\circ}C$

(c) $300^{\circ}F$ = _____ $^{\circ}C$

(d) 60 miles = _____ km

(e) 40 miles per hour = _____ km/h

(f) 40 inches = _____ cm

(g) $20^{\circ}F$ = _____ $^{\circ}C$

(h) $220^{\circ}F$ = _____ $^{\circ}C$

(i) 3000 miles = _____ km

(j) 16 inches = _____ mm

(k) 1 mile = _____ m

(l) $98.6^{\circ}F$ = _____ $^{\circ}C$ (You should know this one exactly.)

(m) 22 inches = _____ mm

(n) 40,000 inches = _____ km

(o) $-20^{\circ}F$ = _____ $^{\circ}C$

- - - - - - - - - - - - - - - -

(a) 15 cm; (b) $20^{\circ}C$; (c) $150^{\circ}C$; (d) 96 km; (e) 64 km/h; (f) 100 cm;
(g) $-5^{\circ}C$; (h) $110^{\circ}C$; (i) 4800 km; (j) 400 mm; (k) 1600 m; (l) $37^{\circ}C$;
(m) 550 mm; (n) 1 km; (o) $-35^{\circ}C$

Weight: Ounces and Pounds to Grams

12. Review your sense of metric weight first. What do these objects
    weigh (approximately)?

(a) A flashlight battery weighs _____ kg.

(b) A straight pin weighs _____ g.

(c) A nickel weighs _____ g.

(d) A six-pack of soda weighs _____ kg.

(e) A can of tuna weighs _____ g.

- - - - - - - - - - - - - - - -

(a) 0.1 kg; (b) 0.1 g; (c) 5 g; (d) 2.5 kg; (e) 250 g

13. When converting from pounds to kilograms, you'll find it helpful to remember that a pound is smaller than a kilogram (1 lb = 0.45 kg). We can then make a fair approximation with this rule:

Take half, and then some.

So, 100 pounds is converted to kilograms, as here:

$\frac{1}{2}$ (100) = 50 – 5 = 45 kg

You can make a very accurate conversion if you "take half" and then let the "then some" be 10 percent of what is left. Let's see how it works out for 60 pounds.

$\frac{1}{2}$ (60) = 30
10% of 30 = 3
60 pounds = 30 – 3 kg, or 27 kg

Practice these conversions by doing them in your head. Remember the rule: Take half and then some. Watch the prefixes. (To convert pounds to grams, convert first to kg; then move the decimal three places to the right.)

(a) 90 lb = _____ kg

(b) 40 lb = _____ kg

(c) 30 lb = _____ kg

(d) 220 lb = _____ kg

(e) 1 lb = _____ g

(f) 6 lb = _____ g

– – – – – – – – – – – – – – – –

(a) 40.5 kg; (b) 18 kg; (c) 13.5 kg; (d) 99 kg; (e) 450 g; (f) 2700 g

Weight: Ounces to Grams

14. To make a rough conversion from ounces to grams, you'll find it easiest to change ounces to pounds (16 oz = 1 lb) and then convert. Your head can only hold so many rules.*
To convert 4 ounces to grams, use 0.25 lb.

4 oz = 0.25 lb
$\frac{1}{2}$ (0.25) = 0.125

_____

* If you do have a particularly good memory, here's a direct ounce-to-gram rule: Multiply by 30 and subtract 10 percent of what is left. So, 100 ounces = 100 × 30 = 3000 – 300 = 2700 grams. (The exact answer is 2835 g.)

Now subtract some—10 percent of what was left.

$$0.125 - 0.013 = 0.11 \text{ kg, or } 110 \text{ g}$$

Try the following conversions. With odd numbers, you can round to an even number, since the conversion is just an approximation.

(a) 8 oz = _____ kg, or _____ g

(b) 12 oz = _____ g

(c) 3 oz = _____ g

(d) 15 oz = _____ g

(e) 6 lb 8 oz = _____ kg, or _____ g

– – – – – – – – – – – – – – – –

(a) 0.225 kg or 225 g; (b) 340 g; (c) 90 g; (d) 450 g; (e) 2.95 kg or 295 g

Review

15. Look over the rules you've learned so far:

- $^\circ$F to $^\circ$C: Take half and subtract 15.

- Inches to centimeters: Double and add half.

- Miles to kilometers: Add 60 percent.

- Pounds to kilograms: Take half and then some.

Review the general rule:

- When converting from a larger to a smaller unit, you get a larger number.

- When converting from a smaller to a larger unit, you get a smaller number.

Now note how the general rule applies for each of the specific conversion rules:

- $^\circ$F are smaller than $^\circ$C. (Between 32°F and 212°F are 180 degrees.)
  Take half and subtract 15.

- Inches are longer than centimeters.
  Double and add half.

- Miles are longer than kilometers.
  Add 60 percent.

- A pound is smaller than a kilogram.
  Take half and then some.

Another set of memory devices that may help you centers around the shapes of the initial letters of the NAS units. Think of the initial letters in pounds and Fahrenheit, the P and the F, as letters cut in half:

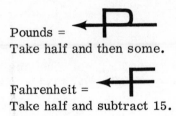

Pounds =
Take half and then some.

Fahrenheit =
Take half and subtract 15.

For the mile and the inch, think of these aids:

Mile = More
Add 60 percent.

Inch = Increase
Double and add half.

Now try these problems. (The arithmetic has been kept easy. Just watch your prefixes!)

(a) $200^{\circ}F = $ _____ $^{\circ}C$

(b) 60 miles = _____ km

(c) 30 inches = _____ cm

(d) 40 pounds = _____ kg

(e) 10 inches = _____ mm

(f) 2 pounds = _____ g

(g) $36^{\circ}F = $ _____ $^{\circ}C$

(h) 120 inches = _____ cm

(i) 4 miles = _____ m

(j) $70^{\circ}F = $ _____ $^{\circ}C$

(k) 900 pounds = _____ kg

(l) 300 miles = _____ km

(m) $0^{\circ}F =$ _____ $^{\circ}C$

(n) 10 miles = _____ km

(o) 2 miles = _____ m

(p) 16 pounds = <u>7.2</u> kg

(q) 40 inches = _____ m

(r) 60 pounds = <u>27</u> kg

(s) 1000 pounds = _____ kg

(t) $-10^{\circ}F =$ _____ $^{\circ}C$

– – – – – – – – – – – – – – – –

(a) $85^{\circ}C$; (b) 96 km; (c) 75 cm; (d) 18 kg; (e) 2500 mm; (f) 900 g;
(g) $3^{\circ}C$; (h) 300 cm; (i) 6400 m; (j) $20^{\circ}C$; (k) 405 kg; (l) 480 km;
(m) $-15^{\circ}C$; (n) 16 km; (o) 3200 m; (p) 7.2 kg; (q) 1 m; (r) 27 kg;
(s) 450 kg; (t) $-20^{\circ}C$

Yards and Quarts to Meters and Liters

16. These last conversions are easy. Let's do quarts to liters first.
It actually takes 1.057 quarts to make a liter. If you treat a quart
as if it were a liter, you make an error of only about 6 percent.
Just remember that a liter is a little larger than a quart:

> Liter is Larger (than a quart)

With yards to meters, it takes exactly 1.094 yards to make a
meter. For rough estimates you can consider yards and meters as
approximately the same since the error is only about 10 percent.
Just remember that a meter is a bit more than a yard:

> Meter is More (than a yard)

Use these rules to make the following conversions:

(a) 10 yards = _____ m

(b) 10 quarts = _____ l

(c) 50 yards = _____ m

(d) 70 pounds = _____ kg

(e) 16 inches = _____ cm

(f) 80 pounds = _____ kg

(g) 10 quarts = _____ l

(h) 90 inches = _____ cm

(i) 50 miles = _____ km

(j) 250°F = _____ °C

(k) 80 miles = _____ km

(l) 20 pounds = _____ kg

(m) 120°F = _____ °C

(n) 44 inches = _____ cm

(o) 18 pounds = _____ kg

(p) 26°F = _____ °C

(q) 100 miles = _____ km

- - - - - - - - - - - - - - - -

(a) 9 m; (b) 9.5 liters; (c) 45 m; (d) 32 kg; (e) 40 cm; (f) 36 kg;
(g) 9.5 liters; (h) 225 cm; (i) 80 km; (j) 110°C; (k) 128 km;
(l) 9 kg; (m) 45°C; (n) 110 cm; (o) 8.1 kg; (p) -2°C; (q) 160 km

B. PRECISION CONVERSIONS

17. The exercises in the remainder of this chapter are designed for
students of science and technology who must work with measurements units regularly and, in doing so, need to use the conversion
tables in the Appendix. If you are one of these students, you don't
need these exercises if you can work the three problems that follow:

(a) $10^{-3} \times 10^{-6} =$ _____

(b) If 1 yard = 9.144×10^{-4} km, what do 10 yards equal?

_____ km

(c) If 1 foot = 3.048×10^{-4} km, what do 1000 feet equal?

_____ km

- - - - - - - - - - - - - - - -

(a) 10^{-9} or 0.000000001; (b) 0.009144 km; (c) 0.3048 km

If you were able to work these problems correctly, you already know
how to do precision conversions. If you had trouble with these problems, go to exercise 18, where you will learn more about how to
work them.

Exponents

18. To use the conversion tables in the Appendix, you need to know how to multiply with exponents.

$$10^3 = 10 \times 10 \times 10 = 1000$$

$10^3 \longleftarrow$ This is an exponent. It tells you the number of times to multiply the number by itself. In this case, multiply the number $\underline{3}$ times.

As you know, 10^2 is equal to 100. Larger numbers can also be expressed as exponents. The exponent indicates the number of zeros.

$$10^3 = 1000 \ (3 \text{ zeros})$$
$$10^5 = 100,000 \ (5 \text{ zeros})$$

We can also express numbers smaller than ten by using exponents. A negative exponent indicates that a number is less than one (i.e., a decimal fraction). But the negative exponent specifies the number of zeros <u>after</u> the decimal point plus one digit (i.e., the number of decimal places).

$$10^{-4} = 0.0001 \ (3 \text{ zeros plus 1 digit}) \text{ or } \frac{1}{10,000}$$

$$10^{-7} = 0.0000001 \ (6 \text{ zeros plus 1 digit}) \text{ or } \frac{1}{10,000,000}$$

Write out the following quantities:

(a) $10^{-5} = $ _____

(b) $10^4 = $ _____

(c) $10^{-3} = $ _____

(d) $10^{-1} = $ _____

(e) $10^1 = $ _____

(f) $10^0 = $ _____

- - - - - - - - - - - - - - - -

(a) 0.00001; (b) 10,000; (c) 0.001; (d) 0.1; (e) 10; (f) 1

19. Now let's go the other way. For a whole number that is a power of 10, write 10 with an exponent equal to the number of zeros.

$$10,000 \text{ (4 zeros)} = 10^4$$
$$100,000,000 \text{ (8 zeros)} = 10^8$$

For a decimal number that is equal to a power of 10, write 10 with a negative exponent equal to the number of decimal places.

$$0.00001 \text{ (5 decimal places)} = 10^{-5}$$
$$0.01 \text{ (2 decimal places)} = 10^{-2}$$

Try a few yourself:

(a) $1,000,000,000 = $ _____

(b) $10,000,000 = $ _____

(c) $0.000000000001 = $ _____

(d) $0.0000001 = $ _____

(e) $1,000,000 = $ _____

(f) $0.000001 = $ _____

- - - - - - - - - - - - - - - -

(a) 10^9; (b) 10^7; (c) 10^{-12}; (d) 10^{-7}; (e) 10^6; (f) 10^{-6}

20. As you see, one of the many conveniences of using exponents to express numbers is that you can describe many decimal positions in a brief space, as here:

$$1,000,000,000,000,000,000 = 10^{18}$$

Imagine having to write out a number like $10^{6,000,000}$!

An even more important convenience of the exponent is that it enables us to simplify multiplication:

Using numbers without exponents

$$1,000,000 \times 1000 = $$

$$
\begin{array}{r}
1,000,000 \\
\times\ 1000 \\
\hline
1,000,000,000
\end{array}
$$

Using numbers with exponents

$$10^6 \times 10^3 = 10^9$$

Notice that all we did to multiply the numbers with exponents was to <u>add</u> the exponents.

Try these:

(a) $10^6 \times 10^9 =$ _____

(b) $10^3 \times 10^5 =$ _____

- - - - - - - - - - - - - - -

(a) 10^{15};   (b) 10^8

21. More complex numbers can also be expressed with exponents.

$$239,000 = 2.39 \times 100,000 \text{ or } 2.39 \times 10^5$$

The number must be broken into two factors. Write the first factor as a power of 10 with the exponent equal to one less than the number of digits. The other factor consists of the first digit expressed as a whole number, with remaining digits expressed as the decimal fraction.

$$1920 = 10^3 \times 1.92$$

Express these numbers as exponents:

(a) $21,000 =$ _____

(b) $467,000 =$ _____

(c) $98,760 =$ _____

(d) $12 =$ _____

- - - - - - - - - - - - - -

(a) 2.1×10^4;   (b) 4.67×10^5;   (c) 9.876×10^4;   (d) 1.2×10^1

22. As you have seen, decimal fractions and reciprocals* can be expressed as negative exponents.

$$0.01 = \frac{1}{10^2} = 10^{-2}$$

Since $1/10^2$ is equal to 0.01, and $1/100$ is equal to 0.01, the negative exponent in 10^{-2} tells us to move the decimal point to the left so that the number of decimal places will be equal to the negative exponent.

$$10^{-2} = {.01} = 0.01 \text{ (Move two places)}$$

$$10^{-5} = {.00001} = 0.00001 \text{ (Move five places)}$$

_____

* The reciprocal of a number is 1 over the number. So, the reciprocal of 9 is $\frac{1}{9}$.

Since numbers with exponents are multiplied by adding the exponents,

$$10^{-2} \times 10^5 = 10^3$$

The same operation can be done with more complicated numbers, as here:

$$0.002 \times 200,000 = 2 \times 10^{-3} \times 2 \times 10^5$$
$$= 4 \times 10^2 \text{ or } 400$$

Multiply these numbers. Convert to exponents first, however.

(a) $0.000006 \times 60,000,000 =$ _____

(b) $0.004 \times 3000 =$ _____

(c) $0.00008 \times 0.0002 =$ _____

(d) $220,000 \times 111,000 =$ _____

- - - - - - - - - - - - - - - -

(a) 3.6×10^2 or 360; (b) 1.2×10^1 or 12; (c) 1.6×10^{-10};
(d) 2.442×10^{10}

23. The Appendix includes conversion tables that can be used to make exact conversions from, say, pounds to kilograms or from kilograms to pounds. Take a moment to look over two or three of these tables. Then we'll show you how to use them, if you're aren't sure that you can.

 Notice that the conversion factors in the tables are expressed with exponents of 10.

$$1 \text{ inch} = 2.540 \times 10^{-2} \text{ meters}$$

Here is how to use a simplified version of the conversion table:

(a) If you know inches, look here. ────────────┐

┌─ (b) If you want to know centimeters, look here. │

Multiply → by Obtain ↓	Meter	Centimeter	Inch	
Meter	1	10^{-2}	2.540×10^{-2}	(c) Find your conversion factor:
Centimeter	10^2	1	2.540	inch =
Inch	3.937×10	3.937×10^{-1}	1	2.540 cm.

To make certain you know how to use the conversion tables, use the sample on the preceding page to answer the following questions:

(a) 1 meter = _____ centimeters

(b) 10 centimeters = _____ inches

(c) 20 inches = _____ centimeters

- - - - - - - - - - - - - - - -

(a) 10^2 cm or 100 cm; (b) 3.937 in.; (c) 50.8 cm

So we come to a closing meter:

Now that you've come to the end of the book
And learned to think metric wherever you look,
We hope by the liters and a million square meters,
The benefits equal the energy it took.

Self-Tests

CHAPTER TWO

1. Write the prefixes for the following multiples:

   (a) 0.001 _____   (d) 0.1 _____

   (b) 0.01 _____   (e) 100 _____

   (c) 10 _____   (f) 1000 _____

CHAPTER THREE

1. Draw lines of these lengths:

   (a) 5 cm   (d) 275 cm

   (b) 190 cm   (e) 87 cm

   (c) 35 cm   (f) 208 cm

2. Give estimated speeds for each of the following kinds of driving:

   (a) City driving _____ km/h

   (b) Country driving _____ km/h

   (c) Thruway speed _____ km/h

   (d) Jogging _____ km/h

   (e) School-zone driving _____ km/h

   (f) Walking _____ km/h

CHAPTER FOUR

1. A milliliter is the size of a box _____ on the side.

2. A liter is the size of a box _____ on the side.

3. A cubic meter is the size of a box _____ on the side.

4. How much space is in a closet 2 meters long, 1 meter wide, and 0.5 meter deep? _____

5. A half-carton of milk holds about _____ liter(s).

CHAPTER FIVE

1. Estimate the weight of each of the following objects:

   (a) A nickel _____ g

   (b) An average-sized man _____ kg

   (c) A flashlight battery _____ g

   (d) A can of tuna _____ g

   (e) A telephone receiver _____ g

   (f) A liter of water _____ g

   (g) A six-pack of soda _____ g

   (h) 5 straight pins _____ g

CHAPTER SIX

1. Give these temperature estimates:

   (a) Water freezes at _____ ^{o}C

   (b) Water boils at _____ ^{o}C

   (c) Room temperature is about _____ ^{o}C

   (d) A hot day in New York would be _____ ^{o}C

   (e) A hot day in the desert is _____ ^{o}C

   (f) Normal body temperature is _____ ^{o}C

   (g) Gold melts at _____ ^{o}C

   (h) Meat broils at _____ ^{o}C

CHAPTER SEVEN

1. What are the metric units for these quantities:

   (a) Force _____

   (b) Pressure _____

   (c) Energy _____

   (d) Power _____

   (e) Large power _____

2. How much downward force does a flashlight battery exert on a weight scale? _____

3. About how much pressure is required to inflate an auto tire?

   _____

4. How much work do you do when you lift a flashlight battery 1 meter?

   _____

5. How much power is required to burn a 100-watt bulb for 10 seconds?

   _____

CHAPTER EIGHT

A. Rule-of-Thumb Conversions

1. Convert these quantities to metric units by using a rule-of-thumb:

   (a) 60 inches = _____ cm

   (b) 10 pounds = _____ kg

   (c) 160°F = _____ °C

   (d) 400°F = _____ °C

   (e) 20 miles = _____ km

B. Precision Conversions

2. Use conversion tables to convert the following quantities to metric units:

(a) 100 gilberts = _____ amps (magnetomotive force)

(b) 1000 Btu = _____ J (energy)

(c) 100 horsepower = _____ W (power)

(d) 10 bushels = _____ l (volume)

ANSWERS TO SELF-TESTS

Chapter Two

1. (a) milli-; (b) centi-; (c) deca-; (d) deci-; (e) hecto-; (f) kilo

Chapter Three

1. To check your answers, measure the lines you drew with a metric ruler.
2. (a) 50 km/h; (b) 75 km/h; (c) 100 km/h; (d) 10 km/h; (e) 25 km/h; (f) 5 km/h

Chapter Four

1. 1 cm
2. 1 dm
3. 1 m
4. 1 m^3
5. 0.5 liter

Chapter Five

1. (a) 5 g; (b) 80 kg; (c) 100 g; (d) 250 g; (e) 300 g; (f) 1 kg; (g) 2500 g; (h) 0.5 g

Chapter Six

1. (a) 0°C; (b) 100°C; (c) 20°C; (d) 30°C; (e) 40°C; (f) 37°C; (g) 1000°C; (h) 250°C

Chapter Seven

1. (a) newton; (b) pascal; (c) joule; (d) watt; (e) kilowatt
2. 1 N
3. 2 bars
4. 1 joule
5. 100 joules per second

Chapter Eight

1. (a) 150 cm; (b) 4.5 kg; (c) 115°C; (d) 200°C; (e) 32 km

2. (a) 79.58 amps; (b) 1.055×10^6 J or $1,055,000$ J; (c) 7.457×10^4 W or $74,570,000$ W; (d) 3.524×10^2 $\underline{l}$ or 352.4 $\underline{l}$

Appendix

CONVERSION FACTOR TABLES*

The following twenty-six tables will facilitate ready interconversion of commonly used units and SI units for the physical quantities listed below:

* Reproduced with permission from SI Units by B. Chiswell and E. C. M. Grigg (Sydney, Australia: John Wiley & Sons Australasia Pty Ltd, 1971).

Notes on Tables:

1. Where there are a large number of conversion factors, the tables have been arbitrarily split into (a), (b), and sometimes (c) sub-tables.

2. The correct SI expression for the physical quantity of any table is marked by an asterisk (*), and is found as the first expression in either the horizontal or vertical columns.

TABLE 1 (a) LENGTH (l)
Small

Multiply→ by↘ Obtain ↓	Metre*	Centi-metre	Micron	Nano-metre	Ang-strom	Fermi	Inch
Metre* (m)	1	10^{-2}	10^{-6}	10^{-9}	10^{-10}	10^{-15}	$2 \cdot 540 \times 10^{-2}$
Centimetre (cm)	10^{2}	1	10^{-4}	10^{-7}	10^{-8}	10^{-13}	$2 \cdot 540$
Micron (μ)	10^{6}	10^{4}	1	10^{-3}	10^{-4}	10^{-9}	$2 \cdot 540 \times 10^{4}$
Nanometre (nm)	10^{9}	10^{7}	10^{3}	1	10^{-1}	10^{-6}	$2 \cdot 540 \times 10^{7}$
Angstrom (Å)	10^{10}	10^{8}	10^{4}	10	1	10^{-5}	$2 \cdot 540 \times 10^{8}$
Fermi	10^{15}	10^{13}	10^{9}	10^{6}	10^{5}	1	$2 \cdot 540 \times 10^{13}$
Inch (in)	$3 \cdot 937 \times 10$	$3 \cdot 937 \times 10^{-1}$	$3 \cdot 937 \times 10^{-5}$	$3 \cdot 937 \times 10^{-8}$	$3 \cdot 937 \times 10^{-9}$	$3 \cdot 937 \times 10^{-14}$	1

TABLE 1 (b) LENGTH (l)
Large

Multiply→ by↘ Obtain ↓	Metre*	Centi-metre	Kilo-metre	Inch	Foot	Yard	Rod, pole or perch	Mile
Metre* (m)	1	10^{-2}	10^3	$2 \cdot 540 \times 10^{-2}$	$3 \cdot 048 \times 10^{-1}$	$9 \cdot 144 \times 10^{-1}$	$5 \cdot 029$	$1 \cdot 609 \times 10^3$
Centimetre (cm)	10^2	1	10^5	$2 \cdot 540$	$3 \cdot 048 \times 10$	$9 \cdot 144 \times 10$	$5 \cdot 029 \times 10^2$	$1 \cdot 609 \times 10^5$
Kilometre (km)	10^{-3}	10^{-5}	1	$2 \cdot 540 \times 10^{-5}$	$3 \cdot 048 \times 10^{-4}$	$9 \cdot 144 \times 10^{-4}$	$5 \cdot 029 \times 10^{-3}$	$1 \cdot 609$
Inch (in)	$3 \cdot 937 \times 10$	$3 \cdot 937 \times 10^{-1}$	$3 \cdot 937 \times 10^4$	1	$1 \cdot 200 \times 10$	$3 \cdot 600 \times 10$	$1 \cdot 980 \times 10^2$	$6 \cdot 336 \times 10^4$
Foot (ft)	$3 \cdot 281$	$3 \cdot 281 \times 10^{-2}$	$3 \cdot 281 \times 10^3$	$8 \cdot 333 \times 10^{-2}$	1	$3 \cdot 000$	$1 \cdot 650 \times 10$	$5 \cdot 280 \times 10^3$
Yard (yd)	$1 \cdot 094$	$1 \cdot 094 \times 10^{-2}$	$1 \cdot 094 \times 10^3$	$2 \cdot 778 \times 10^{-2}$	$3 \cdot 330 \times 10^{-1}$	1	$5 \cdot 500$	$1 \cdot 760 \times 10^3$
Rod, pole or perch	$1 \cdot 988$	$1 \cdot 988 \times 10^{-3}$	$1 \cdot 988 \times 10^2$	$5 \cdot 050 \times 10^{-3}$	$6 \cdot 060 \times 10^{-2}$	$1 \cdot 818 \times 10^{-1}$	1	$3 \cdot 200 \times 10^2$
Mile	$6 \cdot 214 \times 10^{-4}$	$6 \cdot 214 \times 10^{-6}$	$6 \cdot 214 \times 10^{-1}$	$1 \cdot 578 \times 10^{-5}$	$1 \cdot 894 \times 10^{-4}$	$5 \cdot 682 \times 10^{-4}$	$3 \cdot 125 \times 10^{-3}$	1

Notes: 1 English nautical mile = 6 080 feet = 1 853·2 metres.
1 International nautical mile = 6 076 feet = 1 852 metres.
1 chain = 22 yd = 66 ft = 20·12 metres.
1 furlong = 10 chains = 220 ft = 201·2 metres.
1 light year = $3 \cdot 105 \times 10^{16}$ ft = $5 \cdot 879 \times 10^{12}$ miles = $9 \cdot 461 \times 10^{15}$ metres.
1 parsec = $1 \cdot 916 \times 10^{13}$ miles = $3 \cdot 084 \times 10^{16}$ metres.

TABLE 2 (a) AREA (l^2)
Small

Multiply→ by↘ Obtain ↓	Square metre*	Square centimetre	Square millimetre	Square inch	Square foot	Square yard
Square metre* (m²)	1	1×10^4	1×10^6	$6 \cdot 452 \times 10^{-4}$	$9 \cdot 290 \times 10^{-2}$	$8 \cdot 361 \times 10^{-1}$
Square centimetre (cm)²	1×10^{-4}	1	1×10^2	$6 \cdot 452$	$9 \cdot 290 \times 10^2$	$8 \cdot 361 \times 10^3$
Square millimetre (mm)²	1×10^{-6}	1×10^{-2}	1	$6 \cdot 452 \times 10^2$	$9 \cdot 290 \times 10^4$	$8 \cdot 361 \times 10^5$
Square inch (in)²	$1 \cdot 550 \times 10^3$	$1 \cdot 550 \times 10^{-1}$	$1 \cdot 550 \times 10^{-3}$	1	$1 \cdot 440 \times 10^2$	$1 \cdot 296 \times 10^3$
Square foot (ft)²	$1 \cdot 0764 \times 10$	$1 \cdot 0764 \times 10^{-3}$	$1 \cdot 0764 \times 10^{-5}$	$6 \cdot 944 \times 10^{-3}$	1	$9 \cdot 000$
Square yard (yd)²	$1 \cdot 1960$	$1 \cdot 1960 \times 10^{-4}$	$1 \cdot 1960 \times 10^{-6}$	$7 \cdot 692 \times 10^{-4}$	$1 \cdot 111 \times 10^{-1}$	1

TABLE 2 (b) AREA (l^2)
Large

Multiply→ by↘ Obtain ↓	Square metre*	Are	Hectare	Square perch	Square chain	Acre	Square mile
Square metre*	1	1×10^2	1×10^4	$1 \cdot 619 \times 10^4$	$4 \cdot 0469 \times 10^2$	$4 \cdot 0469 \times 10^3$	$2 \cdot 590 \times 10^6$
Are	1×10^{-2}	1	1×10^2	$1 \cdot 619 \times 10^2$	$4 \cdot 0469$	$4 \cdot 0469 \times 10$	$2 \cdot 590 \times 10^4$
Hectare	1×10^{-4}	1×10^{-2}	1	$1 \cdot 619$	$4 \cdot 0469 \times 10^{-2}$	$4 \cdot 0469 \times 10^{-1}$	$2 \cdot 590 \times 10^2$
Square perch	$6 \cdot 173 \times 10^{-5}$	$6 \cdot 173 \times 10^{-3}$	$6 \cdot 173 \times 10^{-1}$	1	$1 \cdot 600 \times 10$	$1 \cdot 600 \times 10^2$	$1 \cdot 024 \times 10^5$
Square chain (ch²)	$2 \cdot 471 \times 10^{-3}$	$2 \cdot 471 \times 10^{-1}$	$2 \cdot 471 \times 10$	$6 \cdot 250 \times 10^{-2}$	1	$1 \cdot 000 \times 10$	$6 \cdot 400 \times 10^3$
Acre	$2 \cdot 471 \times 10^{-4}$	$2 \cdot 471 \times 10^{-2}$	$2 \cdot 471$	$6 \cdot 250 \times 10^{-3}$	$1 \cdot 000 \times 10^{-1}$	1	$6 \cdot 400 \times 10^2$
Square mile	$3 \cdot 861 \times 10^{-7}$	$3 \cdot 861 \times 10^{-5}$	$3 \cdot 861 \times 10^{-3}$	$9 \cdot 804 \times 10^{-6}$	$1 \cdot 563 \times 10^{-4}$	$4 \cdot 096 \times 10^{-3}$	1

TABLE 3 (a) VOLUME (l^3)
Small

Multiply→ by↘ Obtain ↓	Cubic metre*	Cubic decimetre (litre)	Cubic centimetre	Cubic millimetre	Cubic inch	Cubic foot	Cubic yard
Cubic metre* (m^3)	1	$1 \cdot 000 \times 10^{-3}$	$1 \cdot 000 \times 10^{-6}$	$1 \cdot 000 \times 10^{-9}$	$1 \cdot 639 \times 10^{-5}$	$2 \cdot 832 \times 10^{-2}$	$7 \cdot 646 \times 10^{-1}$
Cubic decimetre (dm^3) (litre)[a]	$1 \cdot 000 \times 10^3$	1	$1 \cdot 000 \times 10^{-3}$	$1 \cdot 000 \times 10^{-6}$	$1 \cdot 639 \times 10^{-2}$	$2 \cdot 832 \times 10$	$7 \cdot 646 \times 10^2$
Cubic centimetre (cm^3) (millilitre)[a]	$1 \cdot 000 \times 10^6$	$1 \cdot 000 \times 10^3$	1	$1 \cdot 000 \times 10^{-3}$	$1 \cdot 639 \times 10$	$2 \cdot 832 \times 10^4$	$7 \cdot 646 \times 10^5$
Cubic millimetre (mm^3)	$1 \cdot 000 \times 10^9$	$1 \cdot 000 \times 10^6$	$1 \cdot 000 \times 10^3$	1	$1 \cdot 639 \times 10^4$	$2 \cdot 832 \times 10^7$	$7 \cdot 646 \times 10^8$
Cubic inch (in^3)	$6 \cdot 102 \times 10^4$	$6 \cdot 102 \times 10$	$6 \cdot 102 \times 10^{-2}$	$6 \cdot 102 \times 10^{-5}$	1	$1 \cdot 728 \times 10^3$	$4 \cdot 666 \times 10^4$
Cubic foot (ft^3)	$3 \cdot 531 \times 10$	$3 \cdot 531 \times 10^{-2}$	$3 \cdot 531 \times 10^{-5}$	$3 \cdot 531 \times 10^{-8}$	$5 \cdot 787 \times 10^{-4}$	1	$2 \cdot 700 \times 10$
Cubic yard (yd^3)	$1 \cdot 308$	$1 \cdot 308 \times 10^{-3}$	$1 \cdot 308 \times 10^{-6}$	$1 \cdot 308 \times 10^{-9}$	$2 \cdot 143 \times 10^{-5}$	$3 \cdot 704 \times 10^{-2}$	1

Note: [a] The litre is equivalent to $1 \cdot 000\ 027$ dm^3. Unless extremely accurate values are being measured the litre and dm^3 can be treated as identical as can also cm^3 and millilitre.

TABLE 3 (b) VOLUME (l^3)
Large[a]

Multiply→ by↘ Obtain ↓	Cubic metre*	Litre	U.S. pint	U.K. pint	U.S. gallon	U.K. gallon	U.K. heck
Cubic metre (m³)*	1	$1 \cdot 000 \times 10^{-3}$	$4 \cdot 732 \times 10^{-4}$	$5 \cdot 683 \times 10^{-4}$	$3 \cdot 785 \times 10^{-3}$	$4 \cdot 546 \times 10^{-3}$	$9 \cdot 092 \times 10^{-3}$
Litre[b] (l)	$1 \cdot 000 \times 10^{3}$	1	$4 \cdot 732 \times 10^{-1}$	$5 \cdot 683 \times 10^{-1}$	$3 \cdot 785$	$4 \cdot 546$	$9 \cdot 092$
U.S. pint[c] (pt)	$2 \cdot 113 \times 10^{3}$	$2 \cdot 113$	1	$1 \cdot 201$	$8 \cdot 000$	$9 \cdot 608$	$1 \cdot 922 \times 10$
U.K. pint[c] (pt)	$1 \cdot 760 \times 10^{3}$	$1 \cdot 760$	$8 \cdot 327 \times 10^{-1}$	1	$6 \cdot 662$	$8 \cdot 000$	$1 \cdot 600 \times 10$
U.S. gallon[e] (gal)	$2 \cdot 642 \times 10^{2}$	$2 \cdot 642 \times 10^{-1}$	$1 \cdot 250 \times 10^{-1}$	$1 \cdot 501 \times 10^{-1}$	1	$1 \cdot 201$	$2 \cdot 402$
U.K. gallon[e] (gal)	$2 \cdot 200 \times 10^{2}$	$2 \cdot 200 \times 10^{-1}$	$1 \cdot 041 \times 10^{-1}$	$1 \cdot 250 \times 10^{-1}$	$8 \cdot 327 \times 10^{-1}$	1	$2 \cdot 000$
U.K. heck[d]	$1 \cdot 100 \times 10^{2}$	$1 \cdot 100 \times 10^{-1}$	$5 \cdot 205 \times 10^{-2}$	$6 \cdot 125 \times 10^{-2}$	$4 \cdot 164 \times 10^{-1}$	$5 \cdot 000 \times 10^{-1}$	1

Notes: [a] These volumes are all essentially liquid measures.
[b] The litre actually equals $1 \cdot 000\,027 \times 10^{-3}$ m³.
[c] The U.S. liquid (not dry) pint. In both U.S. and U.K. measures, two pints = 1 quart.
[d] The U.S. heck is also equal to two U.S. gallons.
[e] The firkin = 9 gallons.

TABLE 3 (c) VOLUME (l^3)
Large

Multiply→ by↘ Obtain ↓	Cubic* metre	U.S. fluid ounce	U.K. fluid ounce	U.S. bushel	U.K. bushel	U.S. barrel	U.K. barrel
Cubic metre* (m^3)	1	$2 \cdot 957 \times 10^{-5}$	$2 \cdot 841 \times 10^{-5}$	$3 \cdot 524 \times 10^{-2}$	$3 \cdot 637 \times 10^{-2}$	$1 \cdot 192 \times 10^{-1}$	$1 \cdot 637 \times 10^{-1}$
U.S. fluid ounce[a]	$3 \cdot 378 \times 10^{4}$	1	$1 \cdot 041$	$1 \cdot 190 \times 10^{3}$	$1 \cdot 229 \times 10^{3}$	$4 \cdot 027 \times 10^{3}$	$5 \cdot 530 \times 10^{3}$
U.K. fluid ounce[a]	$3 \cdot 521 \times 10^{4}$	$9 \cdot 615 \times 10^{-1}$	1	$1 \cdot 241 \times 10^{3}$	$1 \cdot 281 \times 10^{3}$	$4 \cdot 197 \times 10^{3}$	$5 \cdot 764 \times 10^{3}$
U.S. bushel[b]	$2 \cdot 838 \times 10$	$8 \cdot 392 \times 10^{-4}$	$8 \cdot 063 \times 10^{-4}$	1	$1 \cdot 032$	$3 \cdot 281$	$4 \cdot 646$
U.K. bushel[b]	$2 \cdot 747 \times 10$	$8 \cdot 123 \times 10^{-4}$	$7 \cdot 804 \times 10^{-4}$	$9 \cdot 690 \times 10^{-1}$	1	$3 \cdot 274$	$4 \cdot 497$
U.S. barrel[c]	$8 \cdot 403$	$2 \cdot 485 \times 10^{-4}$	$2 \cdot 387 \times 10^{-4}$	$3 \cdot 048 \times 10^{-1}$	$3 \cdot 056 \times 10^{-1}$	1	$1 \cdot 376$
U.K. barrel[c]	$6 \cdot 098$	$1 \cdot 803 \times 10^{-4}$	$1 \cdot 732 \times 10^{-4}$	$2 \cdot 149 \times 10^{-1}$	$2 \cdot 218 \times 10^{-1}$	$7 \cdot 269 \times 10^{-1}$	1

Notes: [a] The drachm = $1 \cdot 250 \times 10^{-1}$ of an U.K. fluid ounce while the minim = $2 \cdot 083 \times 10^{-3}$ of a U.K. fluid ounce.
[b] Dry bushels.
[c] Dry barrels—the U.S. liquid barrel is equal to $31 \cdot 5$ U.S. gallons, or $1 \cdot 192 \times 10^{-1}$ m^3, or 42 U.S. gallons of oil.

TABLE 4   ANGLE

Multiply→ by↘ Obtain ↓	Radian*	Revolution	Degree	Minute	Second
Radian* (rad)	1	$6 \cdot 283$	$2 \cdot 262 \times 10^3$	$1 \cdot 357 \times 10^5$	$8 \cdot 143 \times 10^6$
Revolution (rev)	$1 \cdot 592 \times 10^{-1}$	1	$2 \cdot 778 \times 10^{-3}$	$4 \cdot 630 \times 10^{-5}$	$7 \cdot 692 \times 10^{-7}$
Degree (°)	$4 \cdot 425 \times 10^{-4}$	$3 \cdot 600 \times 10^2$	1	$1 \cdot 667 \times 10^{-2}$	$2 \cdot 778 \times 10^{-4}$
Minute	$7 \cdot 353 \times 10^{-6}$	$2 \cdot 160 \times 10^4$	$6 \cdot 000 \times 10$	1	$1 \cdot 667 \times 10^{-2}$
Second	$1 \cdot 229 \times 10^{-7}$	$1 \cdot 296 \times 10^6$	$3 \cdot 600 \times 10^3$	$6 \cdot 000 \times 10$	1

TABLE 5   TIME (t)

Multiply→ by↘ Obtain ↓	Second*	Minute	Hour	Solar day	Sidereal day	Solar year
Second* (s)	1	$6 \cdot 000 \times 10$	$3 \cdot 600 \times 10^2$	$8 \cdot 640 \times 10^4$	$8 \cdot 616 \times 10^4$	$3 \cdot 156 \times 10^7$
Minute (min)	$1 \cdot 667 \times 10^{-2}$	1	$6 \cdot 000 \times 10$	$1 \cdot 440 \times 10^3$	$1 \cdot 436 \times 10^3$	$5 \cdot 259 \times 10^5$
Hour (hr)	$2 \cdot 778 \times 10^{-4}$	$1 \cdot 667 \times 10^{-2}$	1	$2 \cdot 400 \times 10$	$2 \cdot 360 \times 10$	$8 \cdot 765 \times 10^3$
Solar day	$1 \cdot 157 \times 10^{-5}$	$6 \cdot 944 \times 10^{-4}$	$4 \cdot 167 \times 10^{-2}$	1	$9 \cdot 833 \times 10^{-1}$	$3 \cdot 652 \times 10^2$
Sidereal day	$1 \cdot 161 \times 10^{-5}$	$6 \cdot 944 \times 10^{-4}$	$4 \cdot 237 \times 10^2$	$1 \cdot 017$	1	$3 \cdot 591 \times 10^2$
Solar year	$3 \cdot 165 \times 10^{-8}$	$1 \cdot 901 \times 10^{-6}$	$1 \cdot 140 \times 10^{-4}$	$2 \cdot 740 \times 10^{-3}$	$2 \cdot 786 \times 10^{-3}$	1

TABLE 6 VELOCITY (u)

Multiply→ by↘ Obtain↓	Metre/second*	Centimetre/second	Kilometre/second	Kilometre/hour	Foot/second	Mile/hour	Knot
Metre/second*	1	$1 \cdot 000 \times 10^{-2}$	$1 \cdot 000 \times 10^{3}$	$2 \cdot 778 \times 10^{-1}$	$3 \cdot 048 \times 10^{-1}$	$4 \cdot 470 \times 10^{-1}$	$5 \cdot 155 \times 10^{-1}$
Centimetre/second	$1 \cdot 000 \times 10^{2}$	1	$1 \cdot 000 \times 10^{5}$	$2 \cdot 778 \times 10$	$3 \cdot 048 \times 10$	$4 \cdot 470 \times 10$	$5 \cdot 155 \times 10$
Kilometre/second	$1 \cdot 000 \times 10^{-3}$	$1 \cdot 000 \times 10^{-5}$	1	$2 \cdot 778 \times 10^{-4}$	$3 \cdot 048 \times 10^{-4}$	$4 \cdot 470 \times 10^{-4}$	$5 \cdot 155 \times 10^{-4}$
Kilometre/hour	$3 \cdot 600$	$3 \cdot 600 \times 10^{-2}$	$3 \cdot 600 \times 10^{3}$	1	$1 \cdot 097$	$1 \cdot 609$	$1 \cdot 852$
Foot/second	$3 \cdot 281$	$3 \cdot 281 \times 10^{-2}$	$3 \cdot 281 \times 10^{3}$	$9 \cdot 113 \times 10^{-1}$	1	$1 \cdot 467$	$1 \cdot 689$
Mile/hour	$2 \cdot 237$	$2 \cdot 237 \times 10^{-2}$	$2 \cdot 237 \times 10^{3}$	$6 \cdot 214 \times 10^{-1}$	$6 \cdot 818 \times 10^{-1}$	1	$1 \cdot 152$
Knot	$1 \cdot 943$	$1 \cdot 943 \times 10^{-2}$	$1 \cdot 943 \times 10^{3}$	$5 \cdot 396 \times 10^{-1}$	$5 \cdot 921 \times 10^{-1}$	$8 \cdot 684 \times 10^{-1}$	1

TABLE 7 (a) MASS (m)
Small

Multiply→ by↘ Obtain ↓	Kilo-gramme*	Gramme	Grain	Dram	Ounce	Pound	Atomic mass unit
Kilogramme* (kg)	1	$1 \cdot 000 \times 10^{-3}$	$6 \cdot 480 \times 10^{-5}$	$1 \cdot 772 \times 10^{-3}$	$2 \cdot 835 \times 10^{-2}$	$4 \cdot 536 \times 10^{-1}$	$1 \cdot 660 \times 10^{-27}$
Gramme (g)	$1 \cdot 000 \times 10^{3}$	1	$6 \cdot 480 \times 10^{-2}$	$1 \cdot 772$	$2 \cdot 835 \times 10$	$4 \cdot 536 \times 10^{2}$	$1 \cdot 660 \times 10^{-24}$
Grain[a] (gr)	$1 \cdot 543 \times 10^{4}$	$1 \cdot 543 \times 10$	1	$2 \cdot 734 \times 10$	$4 \cdot 375 \times 10^{2}$	$6 \cdot 990 \times 10^{3}$	$2 \cdot 561 \times 10^{-23}$
Dram[b] (dr)	$5 \cdot 644 \times 10^{2}$	$5 \cdot 644 \times 10^{-1}$	$3 \cdot 657 \times 10^{-2}$	1	$1 \cdot 600 \times 10$	$2 \cdot 560 \times 10^{2}$	$9 \cdot 369 \times 10^{-25}$
Ounce[b] (oz)	$3 \cdot 527 \times 10$	$3 \cdot 527 \times 10^{-2}$	$2 \cdot 286 \times 10^{-3}$	$6 \cdot 250 \times 10^{-2}$	1	$1 \cdot 600 \times 10$	$5 \cdot 854 \times 10^{-26}$
Pound[b] (lb)	$2 \cdot 205$	$2 \cdot 205 \times 10^{-3}$	$1 \cdot 429 \times 10^{-4}$	$3 \cdot 906 \times 10^{-3}$	$6 \cdot 250 \times 10^{-2}$	1	$3 \cdot 660 \times 10^{-27}$
Atomic mass unit (amu)	$6 \cdot 024 \times 10^{26}$	$6 \cdot 024 \times 10^{23}$	$3 \cdot 904 \times 10^{19}$	$1 \cdot 607 \times 10^{24}$	$1 \cdot 708 \times 10^{25}$	$2 \cdot 732 \times 10^{26}$	1

Notes: [a] The grain is the mass common to the three Imperial systems of weight, namely avoirdupois, troy and apothecary. Thus

$$1 \text{ grain} = 2 \cdot 083 \times 10^{-3} \text{ oz (troy or apothecary)}$$
$$= 2 \cdot 286 \times 10^{-3} \text{ oz (avoirdupois)}.$$

[b] Dram, ounce and pound are all avoirdupois.

TABLE 7 (b) MASS (m)
Large

Multiply→ by↘ Obtain ↓	Kilo-gramme*	Metric slug	Tonne	Pound	Slug	Hun-dred-weight	U.S. or U.K. ton
Kilogramme* (kg)	1	$9 \cdot 806$	$1 \cdot 000 \times 10^3$	$4 \cdot 536 \times 10^{-1}$	$1 \cdot 459 \times 10$	$5 \cdot 080 \times 10$	$1 \cdot 016 \times 10^3$
Metric slug	$1 \cdot 021 \times 10^{-1}$	1	$1 \cdot 021 \times 10^2$	$4 \cdot 631 \times 10^{-2}$	$1 \cdot 490$	$5 \cdot 187$	$1 \cdot 037 \times 10^2$
Tonne (metric ton)	$1 \cdot 000 \times 10^{-3}$	$9 \cdot 806 \times 10^{-3}$	1	$4 \cdot 536 \times 10^{-4}$	$1 \cdot 459 \times 10^{-2}$	$5 \cdot 080 \times 10^{-2}$	$1 \cdot 016$
Pound (lb) (avoirdupois)	$2 \cdot 205$	$2 \cdot 162 \times 10$	$2 \cdot 205 \times 10^3$	1	$3 \cdot 217 \times 10$	$1 \cdot 120 \times 10^2$	$2 \cdot 240 \times 10^3$
Slug (geepound)	$6 \cdot 854 \times 10^{-2}$	$6 \cdot 721 \times 10^{-1}$	$6 \cdot 854 \times 10$	$3 \cdot 109 \times 10^{-2}$	1	$3 \cdot 482$	$6 \cdot 963 \times 10$
Hundredweight[a] (cwt)	$1 \cdot 968 \times 10^{-2}$	$1 \cdot 930 \times 10^{-1}$	$1 \cdot 968 \times 10$	$8 \cdot 929 \times 10^{-3}$	$2 \cdot 871 \times 10^{-1}$	7	$2 \cdot 000 \times 10$
U.S. or U.K. ton[a]	$9 \cdot 842 \times 10^{-4}$	$9 \cdot 651 \times 10^{-3}$	$9 \cdot 842 \times 10^{-1}$	$4 \cdot 464 \times 10^{-4}$	$1 \cdot 436 \times 10^{-2}$	$5 \cdot 000 \times 10^{-2}$	1

Notes: [a] These are long hundredweights and tons.
The so-called short ton = 20 short cwt.
2 000 lb (avoirdupois).

TABLE 8   FORCE (F)

Multiply→ by↘ Obtain ↓	Newton*	Dyne	Gramme force	Kilo-gramme force	Poundal	Pound force
Newton* (N)	1	$1 \cdot 000 \times 10^{-5}$	$9 \cdot 807 \times 10^{-3}$	$9 \cdot 807$	$1 \cdot 383 \times 10^{-1}$	$4 \cdot 448$
Dyne	$1 \cdot 000 \times 10^{5}$	1	$9 \cdot 807 \times 10^{2}$	$9 \cdot 807 \times 10^{5}$	$1 \cdot 383 \times 10^{4}$	$4 \cdot 448 \times 10^{5}$
Gramme force	$1 \cdot 020 \times 10^{2}$	$1 \cdot 020 \times 10^{-3}$	1	$1 \cdot 000 \times 10^{3}$	$1 \cdot 410 \times 10$	$4 \cdot 534 \times 10^{2}$
Kilogramme force	$1 \cdot 020 \times 10^{-1}$	$1 \cdot 020 \times 10^{-6}$	$1 \cdot 000 \times 10^{-3}$	1	$1 \cdot 410 \times 10^{-2}$	$4 \cdot 534 \times 10^{-1}$
Poundal	$7 \cdot 233$	$7 \cdot 233 \times 10^{-5}$	$7 \cdot 093 \times 10^{-2}$	$7 \cdot 093 \times 10$	1	$3 \cdot 217 \times 10$
Pound force	$2 \cdot 248 \times 10^{-1}$	$2 \cdot 248 \times 10^{-6}$	$2 \cdot 208 \times 10^{-3}$	$2 \cdot 208$	$3 \cdot 108 \times 10^{-2}$	1

TABLE 9 (a) PRESSURE (p)
Small

Multiply→ by↘ Obtain ↓	Newton/ metre²*	Torr	Milli-bar	Pound-force/ ft²	Kg-force/ m²	Poundal/ ft²	Dyne/ cm²
Newton/metre²* $(\text{N m}^{-2})^{cdef}$	1	$1 \cdot 333 \times 10^2$	$1 \cdot 000 \times 10^2$	$4 \cdot 788 \times 10$	$9 \cdot 807$	$1 \cdot 488$	$1 \cdot 000 \times 10^{-1}$
Torr (mm Hg)	$7 \cdot 501 \times 10^{-3}$	1	$7 \cdot 501 \times 10^{-1}$	$3 \cdot 591 \times 10^{-1}$	$7 \cdot 356 \times 10^{-2}$	$1 \cdot 116 \times 10^{-2}$	$7 \cdot 501 \times 10^{-4}$
Millibar (mbar)	$1 \cdot 000 \times 10^{-2}$	$1 \cdot 333$	1	$4 \cdot 788 \times 10^{-1}$	$9 \cdot 807 \times 10^{-2}$	$1 \cdot 488 \times 10^{-2}$	$1 \cdot 000 \times 10^{-3}$
Pound-force/ft² $(\text{lb f ft}^{-2})^{a}$	$2 \cdot 089 \times 10^{-2}$	$2 \cdot 784$	$2 \cdot 089$	1	$2 \cdot 048 \times 10^{-1}$	$3 \cdot 108 \times 10^{-2}$	$2 \cdot 089 \times 10^{-3}$
Kg-force/m² $(\text{Kg f m}^{-2})^{b}$	$1 \cdot 020 \times 10^{-1}$	$1 \cdot 360 \times 10$	$1 \cdot 020 \times 10$	$4 \cdot 882$	1	$1 \cdot 517 \times 10^{-1}$	$1 \cdot 020 \times 10^{-2}$
Poundal/ft² (pdl ft^{-2})	$6 \cdot 720 \times 10^{-1}$	$8 \cdot 954 \times 10$	$6 \cdot 720 \times 10$	$3 \cdot 217 \times 10$	$6 \cdot 588$	1	$6 \cdot 720 \times 10^{-2}$
Dyne/cm² (dyn cm^{-2})	$1 \cdot 000 \times 10$	$1 \cdot 333 \times 10^3$	$1 \cdot 000 \times 10^3$	$4 \cdot 788 \times 10^2$	$9 \cdot 807 \times 10$	$1 \cdot 488 \times 10$	1

Notes: [a] "Pound-force/ft²" is frequently abbreviated to "Pound/ft²".
[b] "Kg-force/m²" is frequently abbreviated to "Kg/m²".
[c] 1 pieze (pz) = 10^3N m^{-2}.
[d] 1 bar = 10^5N m^{-2}.
[e] 1 ton-force/in² = $1 \cdot 544 \times 10^7 \text{N m}^{-2}$.
[f] "Pascal" see page 15 Section 4.

TABLE 9 (b) PRESSURE (p)
Large

Multiply→ by↘ Obtain ↓	Newton/ m²*	Atmo- sphere	Kg- force/ cm²	Pound force/ in² (psi)	Inches Hg	Inches H₂O	Mm Hg
Newton/ metre²*[e] (N m^{-2})[f]	1	$1 \cdot 013 \times 10^5$	$9 \cdot 804 \times 10^4$	$6 \cdot 895 \times 10^3$	$3 \cdot 386 \times 10^3$	$2 \cdot 491 \times 10^2$	$1 \cdot 333 \times 10^2$
Atmosphere (atm)	$9 \cdot 869 \times 10^{-6}$	1	$9 \cdot 678 \times 10^{-1}$	$6 \cdot 804 \times 10^{-2}$	$3 \cdot 342 \times 10^{-2}$	$2 \cdot 458 \times 10^{-3}$	$1 \cdot 316 \times 10^{-3}$
Kg-force/cm² (Kg f cm^{-2})[c]	$1 \cdot 020 \times 10^{-5}$	$1 \cdot 033$	1	$7 \cdot 031 \times 10^{-2}$	$3 \cdot 453 \times 10^{-2}$	$2 \cdot 539 \times 10^{-3}$	$1 \cdot 359 \times 10^{-3}$
Pound-force/in² (lb g in^{-2} or psi)[d]	$1 \cdot 450 \times 10^{-4}$	$1 \cdot 470 \times 10$	$1 \cdot 422 \times 10$	1	$4 \cdot 912 \times 10^{-1}$	$3 \cdot 613 \times 10^{-2}$	$1 \cdot 934 \times 10^{-2}$
Inches Hg[a]	$2 \cdot 953 \times 10^{-4}$	$2 \cdot 992 \times 10$	$2 \cdot 896 \times 10$	$2 \cdot 036$	1	$7 \cdot 355 \times 10^{-2}$	$3 \cdot 937 \times 10^{-2}$
Inches H₂O[b]	$4 \cdot 015 \times 10^{-3}$	$4 \cdot 068 \times 10^2$	$3 \cdot 939 \times 10^2$	$2 \cdot 768 \times 10$	$1 \cdot 360 \times 10$	1	$5 \cdot 354 \times 10^{-1}$
Mm Hg[a] (torr)	$7 \cdot 501 \times 10^{-3}$	$7 \cdot 600 \times 10^2$	$7 \cdot 355 \times 10^2$	$5 \cdot 171 \times 10$	$2 \cdot 540 \times 10$	$1 \cdot 868$	1

Notes: [a] "Inches Hg" and "mm Hg" refers to Hg at 0°C or 273K.
[b] "Inches H₂O" refers to water at 4°C or 277K.
[c] "Kg force cm^{-2}" is frequently abbreviated to "Kg cm^{-2}".
[d] "Pound force in^{-2}" is frequently abbreviated to "pound in^{-2}" or "psi".
[e] "Newton per square metre".
[f] "Pascal" see page 15 Section 4.

TABLE 10 (a) ENERGY (E)
Part 1

Multiply→ by↘ Obtain ↓	Joule*	Kilo-watt-hour	Kilo-calorie	British ther-mal unit	Foot-pound	Foot-poundal	Litre-atmo-sphere	Erg
Joule (J)*a	1	$3 \cdot 600 \times 10^6$	$4 \cdot 186 \times 10^3$	$1 \cdot 055 \times 10^3$	$1 \cdot 356$	$4 \cdot 214 \times 10^{-2}$	$1 \cdot 013 \times 10^2$	$1 \cdot 000 \times 10^{-7}$
Kilowatt-hour (kWh)	$2 \cdot 778 \times 10^{-7}$	1	$1 \cdot 163 \times 10^{-3}$	$2 \cdot 930 \times 10^{-4}$	$3 \cdot 766 \times 10^{-7}$	$1 \cdot 171 \times 10^{-8}$	$2 \cdot 815 \times 10^{-5}$	$2 \cdot 778 \times 10^{-14}$
Kilocalorie (kcal)	$2 \cdot 389 \times 10^{-4}$	$8 \cdot 600 \times 10^2$	1	$2 \cdot 520 \times 10^{-1}$	$3 \cdot 239 \times 10^{-4}$	$1 \cdot 007 \times 10^{-5}$	$2 \cdot 421 \times 10^{-2}$	$2 \cdot 389 \times 10^{-11}$
British thermal unit (BTU)	$9 \cdot 480 \times 10^{-4}$	$3 \cdot 412 \times 10^3$	$3 \cdot 969$	1	$1 \cdot 285 \times 10^{-3}$	$3 \cdot 994 \times 10^{-5}$	$9 \cdot 604 \times 10^{-2}$	$9 \cdot 480 \times 10^{-11}$
Foot-pound (ft-lb)	$7 \cdot 377 \times 10^{-1}$	$2 \cdot 655 \times 10^6$	$3 \cdot 086 \times 10^3$	$7 \cdot 783 \times 10^2$	1	$3 \cdot 108 \times 10^{-2}$	$7 \cdot 474 \times 10$	$7 \cdot 367 \times 10^{-8}$
Foot-poundal (ft-pdl)	$2 \cdot 373 \times 10$	$8 \cdot 539 \times 10^7$	$9 \cdot 929 \times 10^4$	$2 \cdot 502 \times 10^4$	$3 \cdot 217 \times 10$	1	$2 \cdot 405 \times 10^3$	$2 \cdot 373 \times 10^{-6}$
Litre-atmo-sphere (l-atm)	$9 \cdot 871 \times 10^{-3}$	$3 \cdot 550 \times 10^4$	$4 \cdot 129 \times 10$	$1 \cdot 041 \times 10$	$1 \cdot 338 \times 10^{-2}$	$4 \cdot 159 \times 10^{-4}$	1	$9 \cdot 869 \times 10^{-10}$
Erg	$1 \cdot 000 \times 10^7$	$3 \cdot 600 \times 10^{13}$	$4 \cdot 186 \times 10^{10}$	$1 \cdot 055 \times 10^{10}$	$1 \cdot 356 \times 10^7$	$4 \cdot 214 \times 10^5$	$1 \cdot 013 \times 10^9$	1

Note: a 1 Thermie $= 4 \cdot 186 MJ = 4 \cdot 186 \times 10^6$ J.

TABLE 10 (b) ENERGY (E)
Part 2

Multiply→ by↘ Obtain ↓	Joule*	Kilo-gramme	Calorie	Erg	Atomic mass unit	Mev	Elec-tron volt	Centi-metre
Joule (J)*	1	$8 \cdot 987 \times 10^{16}$	4.187	$1 \cdot 000 \times 10^{-7}$	$1 \cdot 492 \times 10^{-10}$	$1 \cdot 602 \times 10^{-13}$	$1 \cdot 602 \times 10^{-19}$	$1 \cdot 986 \times 10^{-23}$
Kilogramme (kg)[a]	$1 \cdot 113 \times 10^{-17}$	1	$4 \cdot 659 \times 10^{-17}$	$1 \cdot 113 \times 10^{-24}$	$1 \cdot 660 \times 10^{-27}$	$1 \cdot 783 \times 10^{-30}$	$1 \cdot 783 \times 10^{-36}$	$2 \cdot 210 \times 10^{-40}$
Calorie (cal)[bcd]	$2 \cdot 389 \times 10^{-1}$	$2 \cdot 147 \times 10^{16}$	1	$2 \cdot 389 \times 10^{-8}$	$3 \cdot 564 \times 10^{-11}$	$3 \cdot 827 \times 10^{-14}$	$3 \cdot 827 \times 10^{-20}$	$4 \cdot 745 \times 10^{-24}$
Erg	$1 \cdot 000 \times 10^{7}$	$8 \cdot 987 \times 10^{23}$	$4 \cdot 187 \times 10^{7}$	1	$1 \cdot 492 \times 10^{-3}$	$1 \cdot 602 \times 10^{-6}$	$1 \cdot 602 \times 10^{-12}$	$1 \cdot 986 \times 10^{-16}$
Atomic mass unit (amu)	$6 \cdot 701 \times 10^{9}$	$6 \cdot 025 \times 10^{26}$	$2 \cdot 807 \times 10^{10}$	$6 \cdot 701 \times 10^{2}$	1	$1 \cdot 074 \times 10^{-3}$	$1 \cdot 074 \times 10^{-9}$	$1 \cdot 331 \times 10^{-13}$
Mev	$6 \cdot 242 \times 10^{12}$	$5 \cdot 610 \times 10^{29}$	$2 \cdot 613 \times 10^{13}$	$6 \cdot 242 \times 10^{5}$	$9 \cdot 315 \times 10^{2}$	1	$1 \cdot 000 \times 10^{-6}$	$1 \cdot 240 \times 10^{-10}$
Electron volt (ev)	$6 \cdot 242 \times 10^{18}$	$5 \cdot 610 \times 10^{35}$	$2 \cdot 613 \times 10^{19}$	$6 \cdot 242 \times 10^{11}$	$9 \cdot 315 \times 10^{8}$	$1 \cdot 000 \times 10^{6}$	1	$1 \cdot 240 \times 10^{-4}$
Centimetre^{-1} (cm^{-1})	$5 \cdot 035 \times 10^{22}$	$4 \cdot 524 \times 10^{39}$	$2 \cdot 108 \times 10^{23}$	$5 \cdot 035 \times 10^{15}$	$7 \cdot 513 \times 10^{12}$	$8 \cdot 066 \times 10^{9}$	$8 \cdot 066 \times 10^{3}$	1

Notes: [a] The conversion factors for kilogrammes and atomic mass used were obtained by use of the mass-energy relationship $E = mc^2$.
  [b] 1 15°C calorie = $4 \cdot 1855$J.
  [c] 1 International Table calorie = $4 \cdot 1868$J.
  [d] 1 thermochemical calorie = $4 \cdot 1840$J.

TABLE 11 (a) POWER (P)
Part 1

Multiply→ by↘ Obtain ↓	Watt*	Erg/ second	Kcal minute	Foot- pound/ second	H.P. (British)	BTU min^{-1}	Kilo- watt
Watt (W)* or Joule s^{-1} (J s^{-1})	1	$1 \cdot 000 \times 10^{-7}$	$6 \cdot 977 \times 10$	$1 \cdot 356$	$7 \cdot 457 \times 10^{2}$	$1 \cdot 758 \times 10$	$1 \cdot 000 \times 10^{3}$
Erg/second	$1 \cdot 000 \times 10^{7}$	1	$6 \cdot 977 \times 10^{8}$	$1 \cdot 356 \times 10^{7}$	$7 \cdot 457 \times 10^{9}$	$1 \cdot 758 \times 10^{8}$	$1 \cdot 000 \times 10^{10}$
Kcal/minute	$1 \cdot 433 \times 10^{-2}$	$1 \cdot 433 \times 10^{-9}$	1	$1 \cdot 943 \times 10^{-2}$	$1 \cdot 069 \times 10$	$2 \cdot 520 \times 10^{-1}$	$1 \cdot 433 \times 10$
Foot-pound/ second^a	$7 \cdot 376 \times 10^{-1}$	$7 \cdot 376 \times 10^{-8}$	$5 \cdot 144 \times 10$	1	$5 \cdot 500 \times 10^{2}$	$1 \cdot 297 \times 10$	$7 \cdot 376 \times 10^{2}$
HP (British)	$1 \cdot 341 \times 10^{-3}$	$1 \cdot 341 \times 10^{-10}$	$9 \cdot 355 \times 10^{-2}$	$1 \cdot 818 \times 10^{-3}$	1	$2 \cdot 357 \times 10^{-2}$	$1 \cdot 341$
BTU/min	$5 \cdot 689 \times 10^{-2}$	$5 \cdot 689 \times 10^{-9}$	$3 \cdot 969$	$7 \cdot 712 \times 10^{-2}$	$4 \cdot 241 \times 10$	1	$5 \cdot 689 \times 10$
Kilowatt (kw)	$1 \cdot 000 \times 10^{-3}$	$1 \cdot 000 \times 10^{-10}$	$6 \cdot 977 \times 10^{-2}$	$1 \cdot 356 \times 10^{-3}$	$7 \cdot 457 \times 10^{-1}$	$1 \cdot 758 \times 10^{-2}$	1

Note: ^{a} Foot-pound/second is an abbreviation for foot-pound-force/second (ft lb f s^{-1}).

TABLE 11 (b) POWER (P)
Part 2

Multiply→ by↘ Obtain ↓	Watt* or J s⁻¹	Calorie/ second	Kilo-gramme metre/ second	Foot-pound/ second	Foot-poun-dal/ second	BTU/ hour	HP (Brit-ish)	HP (Met-ric)
Watt (W)* or Joule s⁻¹ (J s⁻¹)	1	4·187	9·807	1·356	$4·214 \times 10^{-2}$	$2·931 \times 10^{-1}$	$7·457 \times 10^{2}$	$7·355 \times 10^{2}$
Calorie/second (cal s⁻¹)	$2·388 \times 10^{-1}$	1	2·343	$3·239 \times 10^{-1}$	$1·007 \times 10^{-2}$	$6·999 \times 10^{-2}$	$1·782 \times 10^{2}$	$1·757 \times 10^{2}$
Kilogramme-metre/second (kg m s⁻¹)	$1·020 \times 10^{-1}$	$4·268 \times 10^{-1}$	1	$1·383 \times 10^{-1}$	$4·296 \times 10^{-3}$	$2·987 \times 10^{-2}$	$7·604 \times 10$	$7·500 \times 10$
Foot-pound/second[a]	$7·376 \times 10^{-1}$	3·087	7·233	1	$3·108 \times 10^{-2}$	$2·161 \times 10^{-1}$	$5·500 \times 10^{2}$	$5·425 \times 10^{2}$
Foot-poundal/second (Ft-pdl s⁻¹)	$2·373 \times 10$	$9·933 \times 10$	$2·328 \times 10^{2}$	$3·218 \times 10$	1	6·955	$1·770 \times 10^{4}$	$1·746 \times 10^{4}$
BTU/hour	3·413	$1·429 \times 10$	$3·347 \times 10$	4·628	$1·438 \times 10^{-1}$	1	$2·545 \times 10^{3}$	$2·511 \times 10^{3}$
HP (British)	$1·341 \times 10^{-3}$	$5·613 \times 10^{-3}$	$1·315 \times 10^{-2}$	$1·818 \times 10^{-3}$	$5·649 \times 10^{-5}$	$3·928 \times 10^{-4}$	1	$9·863 \times 10^{-1}$
HP (Metric)	$1·360 \times 10^{-3}$	$5·694 \times 10^{-3}$	$1·333 \times 10^{-2}$	$1·845 \times 10^{-3}$	$5·728 \times 10^{-5}$	$3·982 \times 10^{-4}$	1·014	1

Note: [a] Foot-pound/second is an abbreviation for foot-pound force second⁻¹ (ft lb f s⁻¹).

TABLE 12 ELECTRIC CHARGE (Q)

Multiply→ by↘ Obtain ↓	Coulomb (C)*	Ampere-hour	Ab-coulomb	Faraday	Stat-coulomb	Franklin
Coulomb (C)*[a] (ampere-second)	1	$3 \cdot 600 \times 10^3$	$1 \cdot 000 \times 10$	$9 \cdot 649 \times 10^4$	$3 \cdot 336 \times 10^{-10}$	$3 \cdot 336 \times 10^{-10}$
Ampere-hour	$2 \cdot 778 \times 10^{-4}$	1	$2 \cdot 778 \times 10^{-3}$	$2 \cdot 681 \times 10$	$9 \cdot 266 \times 10^{-14}$	$9 \cdot 266 \times 10^{-14}$
Abcoulomb	$1 \cdot 000 \times 10^{-1}$	$3 \cdot 600 \times 10^2$	1	$9 \cdot 649 \times 10^3$	$3 \cdot 336 \times 10^{-11}$	$3 \cdot 336 \times 10^{-11}$
Faraday	$1 \cdot 036 \times 10^{-5}$	$3 \cdot 730 \times 10^{-2}$	$1 \cdot 036 \times 10^{-4}$	1	$3 \cdot 457 \times 10^{-15}$	$3 \cdot 457 \times 10^{-15}$
Statcoulomb	$2 \cdot 998 \times 10^9$	$1 \cdot 079 \times 10^{13}$	$2 \cdot 998 \times 10^{10}$	$2 \cdot 893 \times 10^{14}$	1	$1 \cdot 000$
Franklin (Fr)	$2 \cdot 998 \times 10^9$	$1 \cdot 079 \times 10^{13}$	$2 \cdot 998 \times 10^{10}$	$2 \cdot 893 \times 10^{14}$	$1 \cdot 000$	1

Note: [a] 1 international coulomb = $0 \cdot 999 \, 835$ absolute coulomb.

TABLE 13 ELECTRIC CURRENT (I)

Multiply→ by↘ Obtain ↓	Ampere (A)*	Abampere	Biot	Stat-ampere
Ampere (A)*[a]	1	$1 \cdot 000 \times 10$	$1 \cdot 000 \times 10$	$3 \cdot 336 \times 10^{-10}$
Abampere (ab A)	$1 \cdot 000 \times 10^{-1}$	1	$1 \cdot 000$	$3 \cdot 336 \times 10^{-11}$
Biot (Bi)	$1 \cdot 000 \times 10^{-1}$	$1 \cdot 000$	1	$3 \cdot 336 \times 10^{-11}$
Statampere	$2 \cdot 998 \times 10^9$	$2 \cdot 998 \times 10^{10}$	$2 \cdot 998 \times 10^{10}$	1

Note: [a] 1 international ampere = $0 \cdot 999 \, 835$ absolute ampere.

TABLE 14   ELECTRIC POTENTIAL (V)

Multiply→ by↘ Obtain ↓	Volt (V)*	Statvolt	Millivolt	Microvolt	Abvolt
Volt (V)*[a]	1	$2 \cdot 998 \times 10^2$	$1 \cdot 000 \times 10^{-3}$	$1 \cdot 000 \times 10^{-6}$	$1 \cdot 000 \times 10^{-8}$
Statvolt	$3 \cdot 336 \times 10^{-3}$	1	$3 \cdot 336 \times 10^{-6}$	$3 \cdot 336 \times 10^{-9}$	$3 \cdot 336 \times 10^{-11}$
Millivolt (mV)	$1 \cdot 000 \times 10^3$	$2 \cdot 998 \times 10^5$	1	$1 \cdot 000 \times 10^{-3}$	$1 \cdot 000 \times 10^{-5}$
Microvolt (μV)	$1 \cdot 000 \times 10^6$	$2 \cdot 998 \times 10^8$	$1 \cdot 000 \times 10^3$	1	$1 \cdot 000 \times 10^{-2}$
Abvolt (abV)	$1 \cdot 000 \times 10^8$	$2 \cdot 998 \times 10^{10}$	$1 \cdot 000 \times 10^5$	$1 \cdot 000 \times 10^2$	1

Note: [a] 1 international volt $= 1 \cdot 000\ 330$ absolute volts.

TABLE 15   ELECTRIC FIELD STRENGTH (E)

Multiply→ by↘ Obtain ↓	Volt m^{-1}*	Statvolt cm^{-1}	Volt cm^{-1}	Abvolt cm^{-1}	Volt in^{-1}
Volt metre^{-1} (V m^{-1})*	1	$2 \cdot 998 \times 10^4$	$1 \cdot 000 \times 10^2$	$1 \cdot 000 \times 10^{-6}$	$3 \cdot 937 \times 10$
Statvolt cm^{-1}	$3 \cdot 336 \times 10^{-5}$	1	$3 \cdot 336 \times 10^{-3}$	$3 \cdot 336 \times 10^{-11}$	$1 \cdot 313 \times 10^{-3}$
Volt cm^{-1} (V cm^{-1})	$1 \cdot 000 \times 10^{-2}$	$2 \cdot 998 \times 10^2$	1	$1 \cdot 000 \times 10^{-8}$	$3 \cdot 937 \times 10^{-1}$
Abvolt cm^{-1} (ab V cm^{-1})	$1 \cdot 000 \times 10^6$	$2 \cdot 998 \times 10^{10}$	$1 \cdot 000 \times 10^8$	1	$3 \cdot 937 \times 10^7$
Volt inch^{-1} (V in^{-1})	$2 \cdot 540 \times 10^{-2}$	$7 \cdot 615 \times 10^2$	$2 \cdot 540$	$2 \cdot 540 \times 10^{-8}$	1

TABLE 16   ELECTRIC RESISTANCE (R)

Multiply→ by↘ Obtain ↓	Ohm*	Gigohm	Statohm	Megohm	Abohm
Ohm (Ω)*[a]	1	$1\cdot000 \times 10^9$	$8\cdot988 \times 10^{11}$	$1\cdot000 \times 10^6$	$1\cdot000 \times 10^{-9}$
Gigohm (GΩ)	$1\cdot000 \times 10^{-9}$	1	$8\cdot988 \times 10^2$	$1\cdot000 \times 10^{-3}$	$1\cdot000 \times 10^{-18}$
Statohm	$1\cdot113 \times 10^{-12}$	$1\cdot113 \times 10^{-3}$	1	$1\cdot113 \times 10^{-6}$	$1\cdot113 \times 10^{-21}$
Megohm (MΩ)	$1\cdot000 \times 10^{-6}$	$1\cdot000 \times 10^3$	$8\cdot988 \times 10^5$	1	$1\cdot000 \times 10^{-15}$
Abohm (abΩ)	$1\cdot000 \times 10^9$	$1\cdot000 \times 10^{18}$	$8\cdot988 \times 10^{20}$	$1\cdot000 \times 10^{15}$	1

Note: [a] 1 international ohm = $1\cdot000\,495$ absolute ohms.

TABLE 17   ELECTRIC RESISTIVITY (p)
(Formerly called Specific Resistance)

Multiply→ by↘ Obtain ↓	Ohm-m*	Statohm cm	Ohm-in	Ohm-cm	Abohm- cm
Ohm-m (Ω m)*	1	$8\cdot988 \times 10^9$	$2\cdot540 \times 10^{-2}$	$1\cdot000 \times 10^{-2}$	$1\cdot000 \times 10^{-11}$
Statohm-cm	$1\cdot113 \times 10^{-10}$	1	$2\cdot827 \times 10^{-12}$	$1\cdot113 \times 10^{-12}$	$1\cdot113 \times 10^{-21}$
Ohm-inches (Ω in)	$3\cdot937 \times 10$	$3\cdot539 \times 10^{11}$	1	$3\cdot937 \times 10^{-1}$	$3\cdot937 \times 10^{-10}$
Ohm-cm (Ω cm)	$1\cdot000 \times 10^2$	$8\cdot988 \times 10^{11}$	$2\cdot540$	1	$1\cdot000 \times 10^{-9}$
Abohm-cm (ab Ω cm)	$1\cdot000 \times 10^{11}$	$8\cdot988 \times 10^{20}$	$2\cdot540 \times 10^9$	$1\cdot000 \times 10^9$	1

TABLE 18 ELECTRIC CONDUCTANCE (G)

Multiply→ by↘ Obtain ↓	Siemens (S)*	Abmho	Mho	Micro-siemens	Statmho
Siemens (S)* or Ohm^{-1} (Ω^-)	1	$1 \cdot 000 \times 10^9$	$1 \cdot 000$	$1 \cdot 000 \times 10^{-6}$	$1 \cdot 113 \times 10^{-12}$
Abmho	$1 \cdot 000 \times 10^{-9}$	1	$1 \cdot 000 \times 10^{-9}$	$1 \cdot 000 \times 10^{-15}$	$1 \cdot 113 \times 10^{-21}$
Mho (Ω^{-1})[a]	$1 \cdot 000$	$1 \cdot 000 \times 10^9$	1	$1 \cdot 000 \times 10^{-6}$	$1 \cdot 113 \times 10^{-12}$
Microsiemens (μS)[b] or Megohm^{-1}	$1 \cdot 000 \times 10^6$	$1 \cdot 000 \times 10^{15}$	$1 \cdot 000 \times 10^6$	1	$1 \cdot 113 \times 10^{-6}$
Statmho	$8 \cdot 988 \times 10^{11}$	$8 \cdot 988 \times 10^{20}$	$8 \cdot 988 \times 10^{11}$	$8 \cdot 922 \times 10^5$	1

Notes: [a] International mho = $0 \cdot 999$ 505 mho.
 [b] Microsiemens or megohm^{-1} have been called micromho or gemmho.

TABLE 19 CAPACITANCE (C)

Multiply→ by↘ Obtain ↓	Farad*	Abfarad	Micro-farad	Statfarad	Picofarad
Farad (F)*[a]	1	$1 \cdot 000 \times 10^9$	$1 \cdot 000 \times 10^{-6}$	$1 \cdot 113 \times 10^{-12}$	$1 \cdot 000 \times 10^{-12}$
Abfarad (abF)	$1 \cdot 000 \times 10^{-9}$	1	$1 \cdot 000 \times 10^{-15}$	$1 \cdot 113 \times 10^{-21}$	$1 \cdot 000 \times 10^{-21}$
Microfarad (μF)	$1 \cdot 000 \times 10^6$	$1 \cdot 000 \times 10^{15}$	1	$1 \cdot 113 \times 10^{-6}$	$1 \cdot 000 \times 10^{-6}$
Statfarad	$8 \cdot 988 \times 10^{11}$	$8 \cdot 988 \times 10^{20}$	$8 \cdot 988 \times 10^5$	1	$8 \cdot 988 \times 10^{-1}$
Picofarad (pF) (or "puff")	$1 \cdot 000 \times 10^{12}$	$1 \cdot 000 \times 10^{21}$	$1 \cdot 000 \times 10^6$	$1 \cdot 113$	1

Note: [a] 1 international farad = $0 \cdot 999$ 505 absolute farads.

TABLE 20 INDUCTANCE (L)

Multiply→ by↘ Obtain ↓	Henry*	Stathenry	Milli-henry	Abhenry	Pico-henry
Henry (H)*[a]	1	$8 \cdot 988 \times 10^{11}$	$1 \cdot 000 \times 10^{-3}$	$1 \cdot 000 \times 10^{-9}$	$1 \cdot 000 \times 10^{-12}$
Stathenry	$1 \cdot 113 \times 10^{-12}$	1	$1 \cdot 113 \times 10^{-15}$	$1 \cdot 113 \times 10^{-21}$	$1 \cdot 113 \times 10^{-24}$
Millihenry (mh)	$1 \cdot 000 \times 10^{3}$	$8 \cdot 988 \times 10^{14}$	1	$1 \cdot 000 \times 10^{-6}$	$1 \cdot 000 \times 10^{-9}$
Abhenry (abH)	$1 \cdot 000 \times 10^{9}$	$8 \cdot 988 \times 10^{20}$	$1 \cdot 000 \times 10^{6}$	1	$1 \cdot 000 \times 10^{-3}$
Picohenry (pH)	$1 \cdot 000 \times 10^{12}$	$8 \cdot 988 \times 10^{23}$	$1 \cdot 000 \times 10^{9}$	$1 \cdot 000 \times 10^{3}$	1

Note: [a] 1 international henry = $1 \cdot 000$ 495 absolute henry.
The mic (= 10^{-6} henry) was used by the Royal Navy from 1920 to 1938.

TABLE 21 MAGNETIC FLUX (o)

Multiply→ by↘ Obtain ↓	Weber*	Maxwell	Line	Kiloline
Weber (Wb)*	1	$1 \cdot 000 \times 10^{-8}$	$1 \cdot 000 \times 10^{-8}$	$1 \cdot 000 \times 10^{-5}$
Maxwell (Mx)[a]	$1 \cdot 000 \times 10^{8}$	1	$1 \cdot 000$	$1 \cdot 000 \times 10^{3}$
Line	$1 \cdot 000 \times 10^{8}$	$1 \cdot 000$	1	$1 \cdot 000 \times 10^{3}$
Kiloline	$1 \cdot 000 \times 10^{5}$	$1 \cdot 000 \times 10^{-3}$	$1 \cdot 000 \times 10^{-3}$	1

Note: [a] The promaxwell (= 10^{8} maxwell) was used for a short time after 1930 but has been replaced by the weber.

TABLE 22 MAGNETIC INDUCTION, MAGNETIC FLUX DENSITY (B)

Multiply→ by↘ Obtain ↓	Tesla*	Gauss	Maxwell/ cm^2	Gamma
Tesla (T)* (Wb m^{-2})	1	$1\cdot000$ $\times10^{-4}$	$1\cdot000$ $\times10^{-4}$	$1\cdot000$ $\times10^{-9}$
Gauss (Gs or G) (line cm^{-2})	$1\cdot000$ $\times10^4$	1	$1\cdot000$	$1\cdot000$ $\times10^{-5}$
Maxwell/centimetre^2 (Mx cm^{-2})	$1\cdot000$ $\times10^4$	$1\cdot000$	1	$1\cdot000$ $\times10^{-5}$
Gamma[a]	$1\cdot000$ $\times10^9$	$1\cdot000$ $\times10^5$	$1\cdot000$ $\times10^5$	1

Note: [a] The gamma has been used in geophysics since the beginning of the twentieth century.

TABLE 23 MAGNETIC FIELD STRENGTH (H)

Multiply→ by↘ Obtain ↓	Ampere m^{-1}*	Biot cm^{-1}	Abamp cm^{-1}	Oersted	Gilbert cm^{-1}	Ampere in^{-1}
Ampere/metre (Am^{-1})*[ab]	1	$1\cdot000$ $\times10^3$	$1\cdot000$ $\times10^3$	$7\cdot958$ $\times10$	$7\cdot958$ $\times10$	$3\cdot937$ $\times10$
Biot/centimetre (Bi cm^{-1})	$1\cdot000$ $\times10^{-3}$	1	$1\cdot000$	$7\cdot958$ $\times10^{-2}$	$7\cdot958$ $\times10^{-2}$	$3\cdot937$ $\times10^{-2}$
Abampere/centi- metre (abA cm^{-1})	$1\cdot000$ $\times10^{-3}$	$1\cdot000$	1	$7\cdot958$ $\times10^{-2}$	$7\cdot958$ $\times10^{-2}$	$3\cdot937$ $\times10^{-2}$
Oersted (oe)	$1\cdot257$ $\times10^{-2}$	$1\cdot257$ $\times10$	$1\cdot257$ $\times10$	1	$1\cdot000$	$4\cdot949$ $\times10^{-1}$
Gilbert/centimetre (Gb cm^{-1})	$1\cdot257$ $\times10^{-2}$	$1\cdot257$ $\times10$	$1\cdot257$ $\times10$	$1\cdot000$	1	$4\cdot949$ $\times10^{-1}$
Ampere/inch (A in^{-1})	$2\cdot539$ $\times10^{-2}$	$2\cdot539$ $\times10$	$2\cdot539$ $\times10$	$2\cdot021$	$2\cdot021$	1

Notes: [a] The term "ampere-turn" 'has sometimes been used in this context in place of "ampere".
[b] The term "praoersted" has sometimes been used as a unit equal to 4π ampere-turns metre^{-1}.

TABLE 24 MAGNETOMOTIVE FORCE (F_m)

Multiply→ by↘ Obtain ↓	Ampere*	Gilbert	Abampere
Ampere (A)*[a][b]	1	$7 \cdot 958 \times 10^{-1}$	$1 \cdot 000 \times 10$
Gilbert (Gb)	$1 \cdot 257$	1	$1 \cdot 257 \times 10$
Abampere (abA)	$1 \cdot 000 \times 10^{-1}$	$7 \cdot 958 \times 10^{-2}$	1

Notes: [a] The term "ampere-turn" has sometimes been used in this context in place of "ampere".
[b] The term "pragilbert" has sometimes been used as a unit equal to 4π ampere-turns.

TABLE 25 ILLUMINATION

Multiply→ by↘ Obtain ↓	Lux*	Foot-candle	Lumen/ cm^2	Phot
Lux (lx)* ($lm\ m^{-2}$)	1	$1 \cdot 076 \times 10$	$1 \cdot 000 \times 10^4$	$1 \cdot 000 \times 10^4$
Foot-candle (fc)	$9 \cdot 290 \times 10^{-2}$	1	$9 \cdot 290 \times 10^2$	$9 \cdot 290 \times 10^2$
Lumen/centimetre^2	$1 \cdot 000 \times 10^{-4}$	$1 \cdot 076 \times 10^{-3}$	1	$1 \cdot 000$
Phot	$1 \cdot 000 \times 10^{-4}$	$1 \cdot 076 \times 10^{-3}$	$1 \cdot 000$	1

Note: Nox was used by Germany during World War II as a unit. It is equivalent to 10^{-3} lux.

TABLE 26 LUMINANCE

Multiply → by ↘ Obtain ↓	Candela/ m²* (nit)	Candela/ cm² (stilb)	Lambert	Candela/ ft²	Foot- lambert	Apostilb
Candela/metre² (cd m⁻²)* ᵃ or nit (nt)	1	$1 \cdot 000 \times 10^4$	$3 \cdot 183 \times 10^3$	$1 \cdot 076 \times 10$	$3 \cdot 426$	$3 \cdot 183 \times 10^{-1}$
Candela/centi- metre² (cd cm⁻²) or stilb (sb)	$1 \cdot 000 \times 10^{-4}$	1	$3 \cdot 183 \times 10^{-1}$	$1 \cdot 076 \times 10^{-3}$	$3 \cdot 426 \times 10^{-4}$	$3 \cdot 183 \times 10^{-5}$
Lambert (L) or lumen cm⁻² (lm cm⁻²)	$3 \cdot 142 \times 10^{-4}$	$3 \cdot 142$	1	$3 \cdot 381 \times 10^{-3}$	$1 \cdot 076 \times 10^{-3}$	$1 \cdot 000 \times 10^{-4}$
Candela/foot² (cd ft⁻²)	$9 \cdot 290 \times 10^{-2}$	$9 \cdot 290 \times 10^2$	$2 \cdot 957 \times 10^2$	1	$3 \cdot 183 \times 10^{-1}$	$2 \cdot 957 \times 10^{-2}$
Foot-lambert (ft-L) or equivalent foot candle	$2 \cdot 919 \times 10^{-1}$	$2 \cdot 919 \times 10^3$	$9 \cdot 290 \times 10^2$	$3 \cdot 142$	1	$9 \cdot 290 \times 10^{-2}$
Apostilb (asb)ᵇ or lumen metre⁻² (lm m⁻²)	$3 \cdot 142$	$3 \cdot 142 \times 10^4$	$1 \cdot 000 \times 10^4$	$3 \cdot 381 \times 10$	$1 \cdot 076 \times 10$	1

Notes: [a] Luminous intensity of candela $= 98 \cdot 1\%$ that of international candle.
[b] The skot was used by Germany for a short period. It is equivalent to 10^{-3} apostilb or 10^{-3} lumen per square metre.

SUGGESTIONS FOR FURTHER READING

<u>Books</u>

American National Standards Institute, <u>Rules for the Use of Units of the International System of Units</u> (New York: American National Standards Institute, 1969).

Asimov, Isaac, <u>Realm of Measure</u> (Boston: Houghton-Mifflin, 1960).

Bell, A. E., <u>Mechanical Engineering Science</u> (London: Cassell & Company, 1970).

Burton, William K., <u>Measuring Systems and Standards Organizations</u> (New York: American National Standards Institute, 1966).

Frost, Douglas; Helgren, Fred J.; and Sokol, Louis F., <u>Metric Handbook for Hospitals</u> (Waukegan, Ill.: Metric Association, Inc., 1971).

Helgren, Fred, <u>Metric Supplement to Science and Mathematics</u> (Waukegan, Ill.: Metric Association, Inc., 1973).

<u>Articles</u>

"Commercial Weights and Measures," <u>U.S. Metric Study Interim Report</u>, National Bureau of Standards, July 1971.

"Department of Defense," <u>U.S. Metric Study Interim Report</u>, National Bureau of Standards, June 1971.

"Education," <u>U.S. Metric Study Interim Report</u>, National Bureau of Standards, July 1971.

"Engineering Standards," <u>U.S. Metric Study Interim Report</u>, National Bureau of Standards, July 1971.

"A History of the Metric System Controversy in the United States," <u>U.S. Metric Study Interim Report</u>, National Bureau of Standards, August 1971.

"International Standards," <u>U.S. Metric Study Interim Report</u>, National Bureau of Standards, December 1970.

"International Trade," <u>U.S. Metric Study Interim Report</u>, National Bureau of Standards, July 1971.

"Measurement and the Metric System," <u>Elementary Science Packet No. 2</u>, National Science Teachers Association, Washington, D.C.

"A Metric America: A Decision Whose Time Has Come," National Bureau of Standards Technical News Bulletin, U.S. Dept. of Commerce, National Bureau of Standards, July 1971.

"Metric Precision Measuring Instruments," Beloit Tool Corporation, South Beloit, Illinois.

"Nonmanufacturing Businesses," U.S. Metric Study Interim Report, National Bureau of Standards, July 1971.

"Testimony of Nationally Representative Groups," U.S. Metric Study Interim Report, National Bureau of Standards, July 1971.

"The Consumer," U.S. Metric Study Interim Report, National Bureau of Standards, July 1971.

"The International System of Units (SI)," National Bureau of Standards, 1972.

"The Manufacturing Industry," U.S. Metric Study Interim Report, National Bureau of Standards, July 1971.

"U.S.A. 'Goes Metric,'" Beloit Tool Corporation, South Beloit, Illinois.

Films

"Industry Goes Metric," 16 mm, from British Information Services, 845 Third Ave., New York, New York 10022.

"Keys to Metrication," from British Information Services, 845 Third Ave., New York, New York 10022.

"Learning About Metric Measures," 16 mm, color, from BFA Educational Media, 2211 Michigan Avenue, Santa Monica, California 90404 (elementary-junior high).

"A Metric America: A Decision Whose Time Has Come," 16 mm, color/sound, 35 minutes, November 1971, from National Audiovisual Center (GSA), Washington, D.C. 20409.

"The Metric System," from Coronet Instructional Films, 65 East South Water Street, Chicago, Illinois 60601 (grades 6-9).

Information Sources

Metric Association, Inc., Sugarloaf Star Route, Boulder, Colorado 80302 (publishes a quarterly newsletter for members).

American National Standards Institute, 1430 Broadway, New York, New York 10018 (publishes 5 booklets on metric standards for industry).

U.S. Department of Commerce, National Bureau of Standards, Washington, D.C.

Index

METRIC RULER (in centimeters)

NOTES

NOTES

NOTES